Fluid Mechanics and Its Applications

Founding Editor

René Moreau

Volume 131

Series Editor

André Thess, German Aerospace Center, Institute of Engineering Thermodynamics, Stuttgart, Germany

The purpose of this series is to focus on subjects in which fluid mechanics plays a fundamental role. As well as the more traditional applications of aeronautics, hydraulics, heat and mass transfer etc., books will be published dealing with topics, which are currently in a state of rapid development, such as turbulence, suspensions and multiphase fluids, super and hypersonic flows and numerical modelling techniques. It is a widely held view that it is the interdisciplinary subjects that will receive intense scientific attention, bringing them to the forefront of technological advancement. Fluids have the ability to transport matter and its properties as well as transmit force, therefore fluid mechanics is a subject that is particulary open to cross fertilisation with other sciences and disciplines of engineering. The subject of fluid mechanics will be highly relevant in such domains as chemical, metallurgical, biological and ecological engineering. This series is particularly open to such new multidisciplinary domains. The median level of presentation is the first year graduate student. Some texts are monographs defining the current state of a field; others are accessible to final year undergraduates; but essentially the emphasis is on readability and clarity.

Springer and Professor Thess welcome book ideas from authors. Potential authors who wish to submit a book proposal should contact Dr. Mayra Castro, Senior Editor, Springer Heidelberg, e-mail: mayra.castro@springer.com

Indexed by SCOPUS, EBSCO Discovery Service, OCLC, ProQuest Summon, Google Scholar and SpringerLink

More information about this series at https://link.springer.com/bookseries/5980

S. Harikrishnan · A. D. Dhass ·
Hafiz Muhammad Ali

Thermal Performance of Nanofluids in Miniature Heat Sinks with Conduits

Springer

S. Harikrishnan
Department of Mechanical Engineering
Kings Engineering College
Chennai, Tamil Nadu, India

Hafiz Muhammad Ali
Department of Mechanical Engineering
and Interdisciplinary Research Center for
Renewable Energy and Power Systems
(IRC-REPS)
King Fahd University of Petroleum
and Minerals
Dhahran, Saudi Arabia

A. D. Dhass
Department of Mechanical Engineering
Indus Institute of Technology
and Engineering (IITE)
Indus University
Ahmedabad, Gujarat, India

ISSN 0926-5112 ISSN 2215-0056 (electronic)
Fluid Mechanics and Its Applications
ISBN 978-981-16-7844-8 ISBN 978-981-16-7845-5 (eBook)
https://doi.org/10.1007/978-981-16-7845-5

This Springer imprint is published by the registered company Springer Nature Singapore Pte Ltd.
The registered company address is: 152 Beach Road, #21-01/04 Gateway East, Singapore 189721, Singapore

Preface

Heat dissipation requirements have drawn much interests in the last two decades as a result of rapid advancements in a wide range of technologies, particularly in miniature devices. High-performance heat sinks are in demand to ensure the quality and long life of these devices. The demand for an extremely high flow velocity, as well as the related noise and vibration difficulties, has limited the use of forced convection cooling. Owing to the potential of heat carrying capability, the microchannel heat sink appears to be the most reliable cooling method.

It gives us immense pleasure to introduce the book titled *Thermal Performance of Nanofluids in Miniature Heat Sinks with Conduits*, and it comprises five chapters, covering from fundamentals of miniature heat sink using nanofluids to applications. The book starts with types of nanofluids, preparation techniques, thermophysical properties of the nanofluids and types of heat sinks. Flow characteristics of the nanofluids in the miniature heat sinks are discussed with respect to different flow patterns. Next, the modification of design made in miniature heat sink and fabrication process of various conduit miniature heat sinks are included. Then, heat transfer characteristics of the various conduits carrying nanofluids are elaborated. Lastly, various applications of the miniature heat sinks are discussed in detail.

This book could be beneficial to the graduates, research scholars and researchers as it could cater their needs. Suggestions and comments from the readers are highly appreciated to improve the quality of the contents comprised in the book.

Chennai, India
Ahmedabad, India
Dhahran, Saudi Arabia

S. Harikrishnan
A. D. Dhass
Hafiz Muhammad Ali

Acknowledgement Hafiz Muhammad Ali acknowledges the support of King Fahd University of Petroleum and Minerals, Dhahran 31261, Saudi Arabia.

Contents

Chapter 1
Nanofluids

Abstract Nanofluids are colloidal suspensions of nanoparticles with sizes ranging from one to one hundred nanometers in size. Nanofluids have been the subject of momentous study for the last fifteen years, owing to their increased thermal properties. In engineering equipment, the low thermal conductivity of conventional heat transfer fluid has shown to be a significant hindrance to enhancing the performance and compactness of the equipment. The use of tiny particles in nanoscale dispersed in conventional heat transfer fluids can overcome these challenges. Herein, different synthesis techniques of nanoparticles are discussed in detail. There are two methods employed for the preparation of nanofluids and their significance is included. Ultrasonication is a critical step in the preparation process because of its role in stabilizing and enhancing thermophysical characteristics. Further, the effect of nanoparticles on the viscosity of the nanofluids is included as the enhancement of the viscosity decelerates the heat transfer.

Contents

This chapter comprises the introduction of nanofluids and miniature heat sink, types of nanofluids and heat sinks, preparation of nanofluids, characterization of nanofluids and thermo physical properties of nanofluids.

1.1 Introduction

1.1.1 Nanofluid

Fluids are widely used as hot carriers in heat transportation systems. As far as the applications of heat transfer fluids are concerned, hydrothermal heating and cooling

S. Harikrishnan et al., *Thermal Performance of Nanofluids in Miniature Heat Sinks with Conduits*, Fluid Mechanics and Its Applications 131,
https://doi.org/10.1007/978-981-16-7845-5_1

systems, transport cooling systems, petrochemical heating and cooling systems, textiles, etc. are widely used. The mixture of conventional base liquid and measured nanometer partitions is known as Nanofluid. Using nanofluids, low thermal conductivity of base fluids such as air, ethylene glycol and propylene glycol, oil can be replaced. Nanofluids can be used for manufacturing, metal working related uses, heating, ventilation, storage, attractive medicine and a few more, due to their captivating physical characteristics [1]. Many researchers have suggested that one of the main requirements for enhancing the heat transfer is the effect of Brownian motion. Brown's movement is induced by small distortions of the fluid formulation of temperature and velocity.

Heat exchangers are employed to facilitate the heat exchanging process between two fluids at different temperatures. Nanofluids used in the heat exchangers are utilized as coolants in automotive and nuclear reactor heat transfer medium. If nanoparticles are suspended in the base fluids ensuring uniform dispersion, then it can be called as Nanofluids. The sizes of the nanoparticles are typically ranging from 1 to 100 nm however, their lengths and shapes can be slightly different [2, 3].

Nanoparticles are 1000 times higher in surface area for a given mass or volume than normal particles and thus, they have higher suspension stability, improved thermal transport capability with less particle agglomeration and flexible properties. Furthermore, high thermal conductivity, small pressure drops and mechanical degradation make nanofluids highly functional for various industrial applications including microelectronics, transport, space, biomedical, nuclear, solar and cooling systems [4].

A high rate of heat transfer is being pursued for increased requirements for personal computer CPUs and laptop for processing and storage requirements. The supply of different geometries of nanofluids to a heat pipe is a safer choice. The use of nanofluids is a viable choice to increase the engine's cooling rate and reduce the complexity of the engine. In particular, improved heat transfer rates are contributing to the production of lightweight and compact radiators for heavy-duty engines in particular. More specific viscosity correlations are given by the nanofluid flow analysis due to the effect of particle length, aggregation, sonication time and pH, along with particle concentration [5].

Usually, nanofluids have greater heat transport properties than conventional liquids, witnessed through different experimental and theoretical tests. For example, adding 0.3% volume of copper nanoparticles with ethylene glycol, it increased the thermal conductivity by 40%. Dispersion of nanoparticles in the base fluid improves not just the thermal conductivity but also the convective heat transfer values, which also enhances the temperature diffusion due to higher heat transfer range. The engine cooling system using nanofluid is the latest foundation for technological integration and growth. Nanofluids are ideal coolants owing to their high thermal diffusiveness and a fast response to thermal changes. In contrast with pure fluid, the presence of nanoparticles in nanofluids helps to improve the mixing flow and thermal conductivity [6].

1.2 Miniature Heat Sink

High-heat electronic devices produce heat in general and then turn to cool through passive heat exchangers. In an integral and interrupted pin shape, these devices are made. The need for miniature heat sink was primarily due to the need for a smaller scale and a lower weight in many industrial applications. As heat exchangers, the enhancement of the thermal efficiency of micro and mini-channels has been conducive to their use. The classification of miniature heat sink on basis of heat and fluid flow is shown in Fig. 1.1.

In addition, different geometries of the micro channel, such as circular, rectangular, triangular, and trapezoidal designs, are used to increase the heat transfer rate (Fig. 1.2). To build heat sinks for the micro-channel, they have used a variety of materials, such as copper, aluminum and silicon. In order to overcome the electronic chip link weaknesses, a combination of two materials was also suggested.

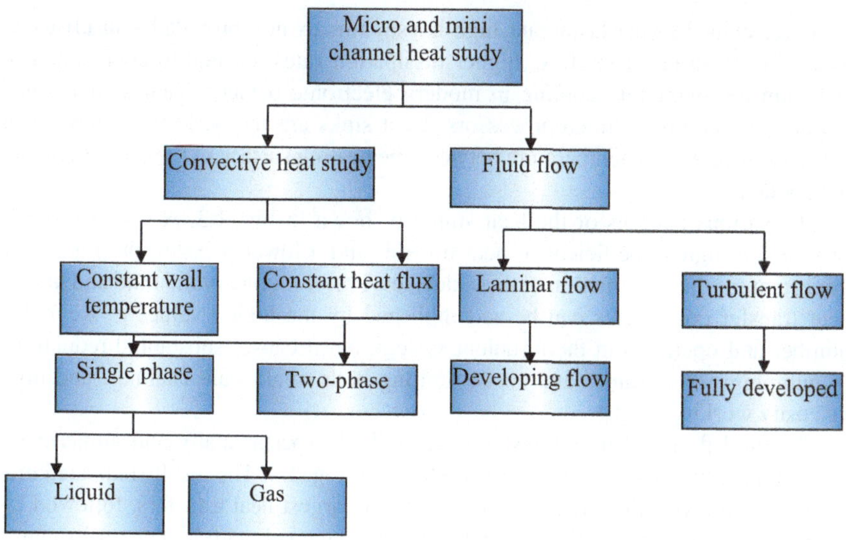

Fig. 1.1 Classification of miniature heat sink based on heat and fluid flow [7]

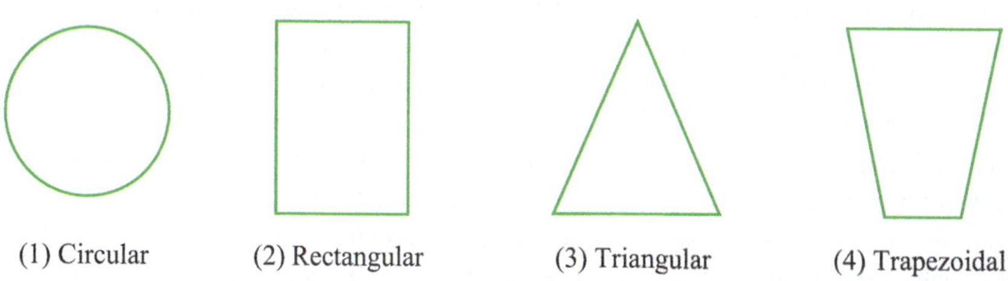

Fig. 1.2 Different type of micro channel fluid flow [8]

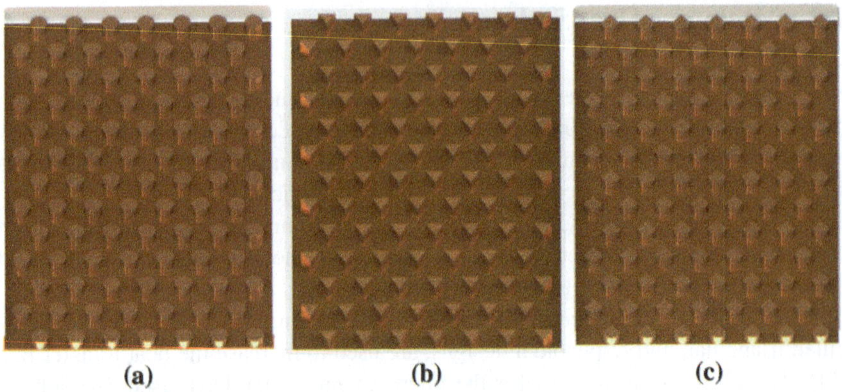

Fig. 1.3 Various patterns of the heat sink with: **a** circular pin fins, **b** triangular pin fins, **c** drop-shaped pin fins [9]

To reject the heat at a faster rate, high-hot streams are now provided with advanced electronic facilities. Therefore, higher dissipation rates are highly appreciated to ascertain the maximum cooling in modern electronic devices, such as integrated circuits and computer microprocessors. Heat sinks are the ways to disperse heat from the difference in temperature between the heating surface of the refrigerant and the coolant.

The various patterns of the heat sink are shown in Fig. 1.3. A strong mixture of fluids, a higher coefficient of heat transfer and a lower pressure drop are very important for the construction of these devices. Profer mixing of the fluid and higher heat transfer coefficients can be accomplished by the higher values of Reynolds number and operation in the turbulent system, but it causes substantial reductions in pumping power values and costs, limiting both small-scale channels and high viscosity working fluids.

The fluid flow and heat transfer were studied experimentally and numerically and compared by serpentine patterns is shown in Fig. 1.4. The results indicated that single-wheel serpentine channel was one of the largest heat transfers, followed by double microchannel and three-wheel serpentine channel, with higher performance from a heat transfer point of view than Spectral Relaxation Method (SRM). The theory was that the re-circulation flow caused by the bent region of the hydrodynamic and thermal boundary layers was skewed and reduced the flow from fully forming. The increase in heat transfer value was accompanied by the increase in pressure reduction. For the determination of the hydrothermal efficiency of serpentine heat sinks, the Performance Evaluation Criteria (PEC) value for the same pumping power was used. The thermal efficiency of the Single Path Serpentine Microchannel (SPSM) heat sink was greatly lowered in order to achieve higher PEC levels with more pump strength. As the pressure decreases dramatically, due to an increase in flow rate, the PEC values decreases at flow rates in all cases above 0.2 l/min.

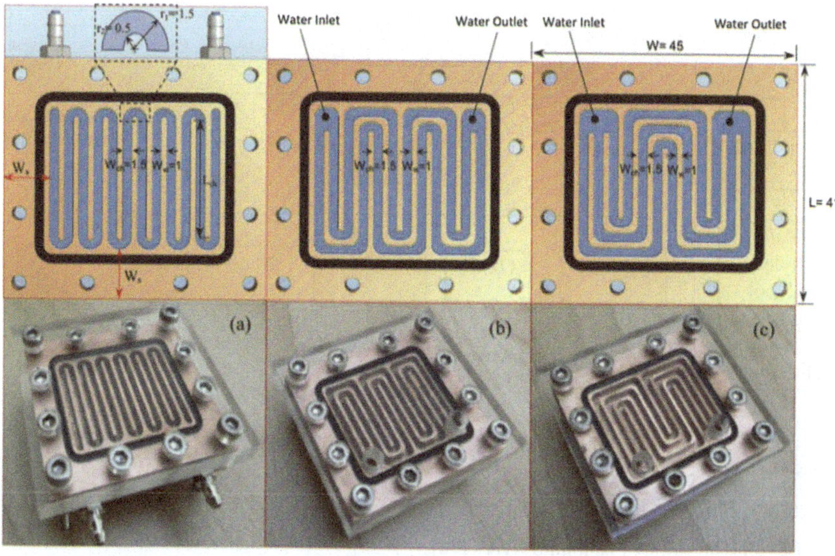

Fig. 1.4 Fluid flow and heat transfer serpentine patterns [9]

1.3 Types of Nanofluids

1.3.1 Base Fluid

Nanofluid flow is characterized as a better model by the two-phase process. In fact, the slip velocity may not be zero due to a variety of variables such as the magnitude, friction between the fluid and the solid. The different types of nanofluids are shown in Fig. 1.5. In a micro-channel under forced laminar convection, the two-phase Eulerian Cu-water nanofluid numeric model was developed. The results show that, although the pressure drop is only marginally higher, the increase in Reynolds number and the volume of nanoparticles and the decrease in the diameters of the nanoparticles are increasing. The thermophysical properties of basefluids is given in Table 1.1.

1.3.2 Metallic Nanoparticles

Metallic NPs can be an attractive way to speed up NP screening for metal and metal oxides in different environmental conditions. However, for grouping with regard to the rate of dissolution, a more thorough documentation of certain dissolution rate constants, along with detailed data on particle characteristics and their interaction with material properties and solvents (particle size distribution, surface area, surface composition and characteristics, ion resistance, complexing, etc.) will be appropriate.

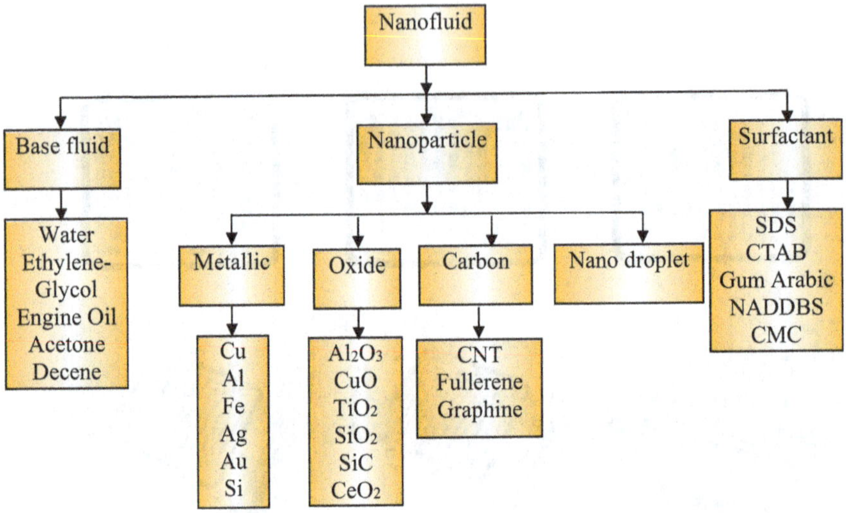

Fig. 1.5 Classification of nanofluids types [10]

Table 1.1 Thermophysical properties of basefluids [11]

Properties	Water	Ethylene glycol	Engine oil	Acetone	Decane
Fluid density ρ (kg/m^3)	998.203	1114.4	884.1	784	0.730
Dynamic viscosity μ (Nm/s)	0.00101	0.0157	0.486	0.000316	0.000859
Thermal conductivity k (W/mK)	0.613	0.252	0.145	0.104002	0.147
Specific heat Cp (kJ/kgK)	4182.2	2415	1909	124.5	300.83

The introduction of nanoparticles greatly increases optical absorption as a result of the electrofield containment around metal nanoparticles [12]. Photographic view of various metal nanoparticles and their anti-bacterial mechanisms are shown in Fig. 1.6.

The interaction of metallic particle dynamics with the corresponding volume also explains its variation. It is the electron-phonon interaction that at the microscopic stage blends the two baths. A numerous researchers, metallic nanoparticles (Fig. 1.6) are extensively used in biomedical and engineering sciences. Their immense nanotechnology ability makes them a focal point of interest. These materials can now be synthesized and modified with various functional chemical groups to be combined with antibodies and medicines of interest to open up a wide range of possible applications [14].

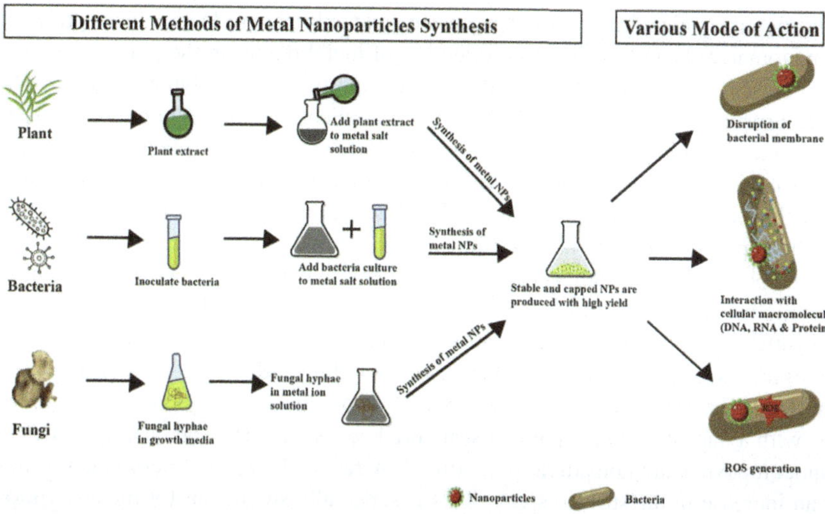

Fig. 1.6 Pictorial representation of metallic nanoparticles from plants, bacteria and fungi for green synthesis [13]

1.3.3 Oxide Nanoparticles

The metal oxides nanoparticles may be of noteworthy physical or chemical properties because of their low size and their high density of corners (or) edge surface sites. The size of the particle should affect three main categories of basic characteristics in any material. It involves the structural features, the parameters of the cells and lattice symmetry. Bulk oxides are generally stable, resilient, and crystallographically well-defined systems. The properties of nanophase structures in modern nanotechnology are their varied nonlinear visual properties, greater ductility at higher temperatures than coarse ceramics, cold welding properties, super-paramagnetic comportment, special catalytic properties, sensitivity and select operations [15].

During different process times (12 and 24 h), nanocrystalline CeO_2 particles with a cubic fluorescent glass structure were synthesized, using them as photoelectrode materials in Dye-Sensitized Solar Cells (DSSC). Three crystalline forms of TiO_2 naturally occur: (1) (tetragonal), (2) (tetragonal) and (3) (orthorhombic). Due to the high charge transport, anatase is the most commonly used phase for solar cell applications, while rutile is used in interference applications due to its high degree of refraction and brookite is difficult for preparation [16].

The key advantages of metal oxide nanoparticles are: (i) structural changes that allow the symmetry of the lattice and cell parameters to be adjusted; (ii) changes in electron due to the effect of quantum containment; and (iii) changes in surface characteristics that lead to a substantial increase in the band gap that affects the conductivity and chemical activity of nanoparticles. Another essential role is biocompatibility with enzymes that lead to selectively sensing biomolecules. Various methods, including solidification and melting, sol-gel processing and polymer production in the presence of nanomaterials and the in-situ production of metal oxides in the polymer matrix, can be used to insert modified metal oxide nanoparticles into polymer matrices. The more prospective and intensively researched applications for a number of catalytic processes, including traditional catalysis of chemical reactions, photocatalysis, and electrocatalysis, are LA (Laser Ablation) oxide nanoparticles. Some oxide nanoparticles have, in addition to catalysis, special electronic and magnetic properties that can be applied in certain places [16].

The physical and chemical properties of metal oxidation nanoparticles have been exciting because of their high density and small scale, and so their different aspects of synthesis, properties and applications are extremely desirable. The homogenous distributions of metal oxide nanoparticles including a decrease in polymer-metal interface size are vital for the development of polymer-metal oxide nanocomposites with a nanostructure-separated structure because the physical characteristics of nanocomposites are considerably modified. A reduced size of material also results in an increase in the surface space. This is especially significant for the adsorption process. Further binding sites are accessible on the surface of the nanomaterials used in the adsorption processes [17].

1.3.4 Carbon Nanoparticles

Almost all types of nanofillers have been used in the manufacture of nanocomposites in recent years: organoclays, CB (carbon black), smoked and colloid silica, CNTs, cellulose whiskers, metal oxide, etc. Carbon nanoparticles (CNPs) are 10 nm in size as an outstanding category of recently developed quasi-spheric carbonate nanomaterials. Figure 1.7 shows the CNP features that are typically useful. In the areas of medicine, energy, electronics, physics, mechanics, materials science, environment, biotechnology and agriculture, nanotechnologies clearly have a promising potential to play a profound role.

In the last two decades, nanostructured allotropic types of carbon have been investigated due to their unusual hybridization and their susceptibility to disruption during synthesis to enable the finest manipulation of the material properties (shown in Fig. 1.8). In different hybridising states, each of which has distinctive properties, carbon can be found in particular.

Preparing it cost-effectively and efficiently can be accomplished with a range of different protocols including microwave radiation, hydrothermal carbonization and pyrolysis. Different instruments are employed to trace the changes in the result of carbon nanoproducts in the hybrid nanoproducts. Due to their unusual combinations (for instantaneous combinations of chemical and physical properties, for example thermal, electrical, high-mechanical and optical resistance), carbon-based nanomaterials become important. CNPs are inert chemically and less toxic, so they are preferred for the bioimaging of traditional quantum dots, nanocarbons and organic dyes and for other treatments. CNT is composed of the sp^2 Carbon Hybridization

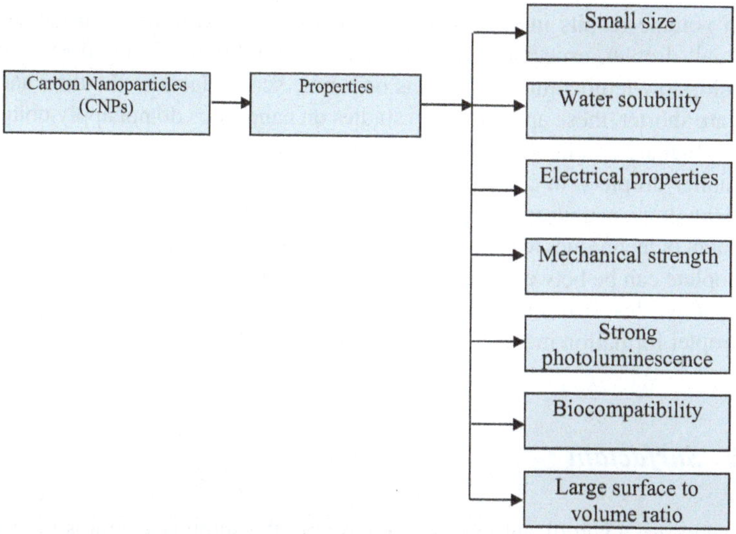

Fig. 1.7 Important properties of CNPs [18]

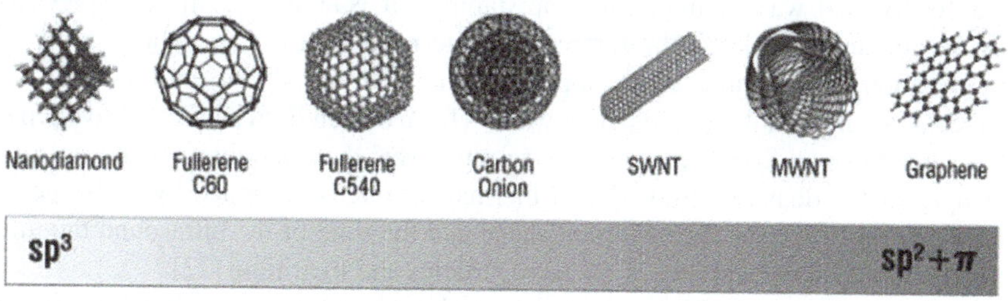

Fig. 1.8 Nanostructured allotropic types of carbon [19]

Network, a graph-like hexagonal net. A remarkable features of modern CNT structures are high chemical stability, high conductivity and high mechanical properties [20].

1.3.5 Nanodroplet

In next generation, coalescence-induced nanodroplets could enhance processor-chip cooling systems and self-cleaning surfaces via passive processes that require little or no input of electricity. The interest in crystallization of nanodroplets was especially high because the freezing temperature observed significantly reduces by 202 kg for the nanodroplets with a radius of 3.2 nm compared with bulk water. Surface-effects usually reduce the temperature of a small system compared to bulk; a smaller system volume results in less nucleation and the high volume-to-surfactant ratio of a small droplet, especially in experimental analysis, allows rapid refrigeration to develop low-temperature conditions on a time scale. Because the time and time scales are shorter, these approaches to studies on nanodrops do not apply ubiquitous artefacts that are crucial to a number of technical tasks, including the use of smaller and smaller droplets, in controlled generation and handling. In biotechnology and nanotechnology, nanodroplets can be widely used. However, these applications need equipment or techniques to ensure stable nanodroplet output at high levels. A charged nanotroplete can be between 1 and several ten nanometers in diameter, even though in gas ions it is less than 1 nm. In accordance with Thomson's theory of ion induction, nanodroplet formation mechanisms can be identified [21].

1.3.6 Surfactant

Surfactants are a significant means of improving the stability of fluids in nanoparticles. The functionality of surfactants at high temperatures in particular, is also of high significance. The integration of dispersants in two-phase systems is a quick and economical way of improving the stability of Nanofluids. Small dispersants can significantly influence the surface characteristics of a system. The stability of the isolation of nanofluids and nanoparticles and their attractions until deposits are solved at the fluid tank or on pipeline walls. This research used and tested five separate surfactants at different levels. The results reveal that there have been no major changes in the density, viscosity and thermal conductivity, added by a change in tensile level. However, the surfactant shape and the time of the ultrasound mixture greatly influence the stability of the nanoparticles and their time [22].

The suspension of solid particles from liquids is common knowledge and contributes to industrial systems for the transfer of heat, lubricant and magnetic fluids. Due to its wide application in heat transfer, microchip cooling, medical supply, and increased oil recovery the nanometer-based additives and base fluids have attracted a great deal of attention as engineering material among these technologies. The addition of surfactants may have a direct impact on the dispersion of nanofluids and thermodynamic properties, with surfactants having few thermal conductivity models [23].

The addition of nanoparticles to the base fluid is an improvement of the heat transfer rate with the increase in the overall Nusselt number at the maximum volume level. The stability improvement is noted in the zeta potential value improvement. The thermophysical properties of carbon-nanotube are determined in the presence of three different surfactants: Sodium Dodecyl Sulphate (SDS), Cetyltrimethylammonium Bromide (CTAB) and Gum Arabic (GA). The surface voltage of nanofluid with increased temperature and SDS has been found to decrease. With increasing surfactant levels and an increased rate of under constant proportions at higher temperatures, viscosity is decreased. The density decreased with increasing temperatures and virtually no additional surfactants.

The main issue is the long-term stability of nanofuids for industrial use. The addition of surfactants is seen as the most simple and economical way to increase the stability of nanofuels. The viscosity of these nanofluid is influenced by nanoparticle size, volume and temperature. It is proposed to stabilize nanofluids with certain chemical or physical techniques. The addition of surfactant is considered a common technique for sedimentary prevention. The volume of surfactant has the right range in order to ensure adequate coating, induce electrostatic repulsion and compensate for Vander Waals forces of interactions. Synthesis and preparation methods for nanofluids are of significance for stability and influence the characteristics of thermal transport of nanofluids. In the field of nanofluid stability, pH is also critical in addition to the filling fraction and the working temperature [24].

1.4 Types of Heat Sink

Continuous electronic device development and operation contribute to the heat generation of these devices. The different types of heat sinks are shown in Fig. 1.9. Several studies take into account water and nanofluid microchannel heat sinks for thermal hydraulic capability [25].

1.5 Based on Manufacturing Methods

Heat sink can be classified based on manufacturing methods that are of familiar in day-to-day applications, so that which is not discussed much on it.

Extruded heat sinks are very cost-effective, but have restricted design flexibility without secondary work, such as CNC machining (Fig. 1.10).

1.5.1 Active Heat Sink

Heat sink components with an electrical device such as a ventilator or a blower are active heat sinks near the heat exchange fluid. This involves heat sinks based on the fluid pumped from a heat source for the extraction of latent heat. Since active thermal sinks are powered by forced air in the Fin Region, heat sink design is smaller and lighter [26].

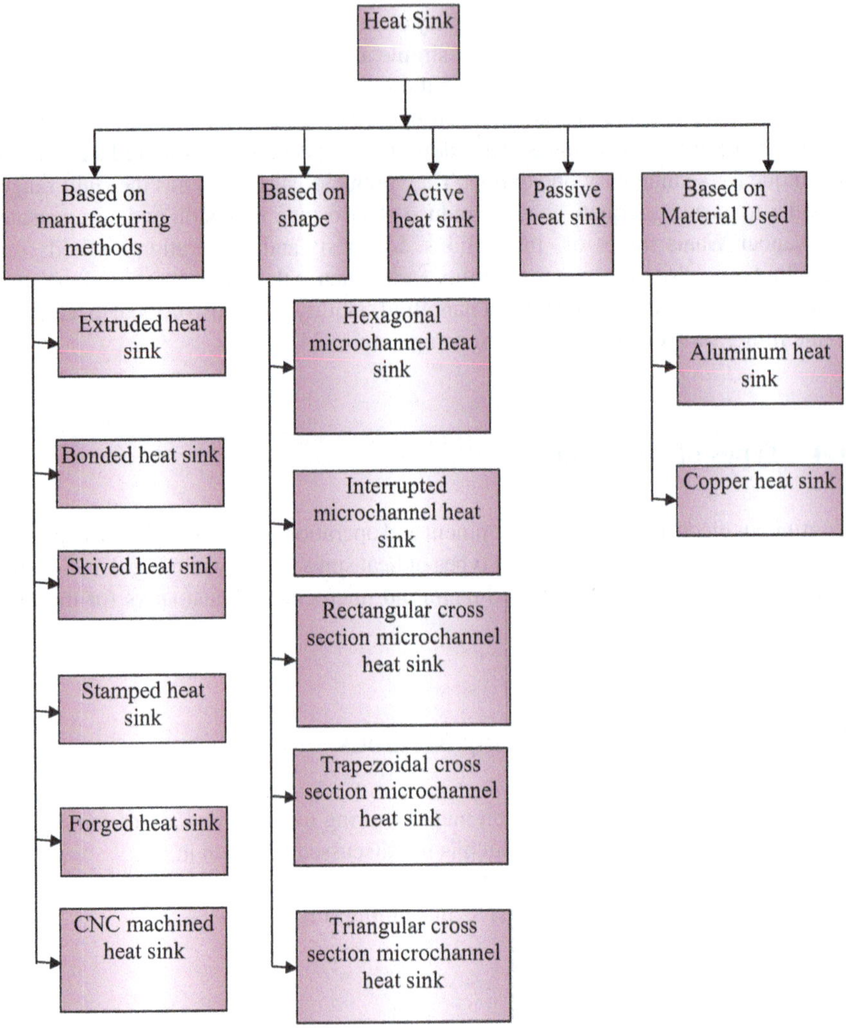

Fig. 1.9 Types of heat sink [26]

1.6 Based on Shape

For several channel formations such as zigzag microchannels, curved and stages, the channel type effect on the Microchannel Heatsink (MCHS) performance is investigated and contrasted with direct and wavy channels. The main step in the design of a heat sink is a channel shape option. The geometry of the channel was then modified to increase the heat transfer range without increasing substantial pumping capacity. Compared to other channels, triangular, trapezoidal, hexagonal, circular, rhombus and rectangular transversal channels have also been defined extensively.

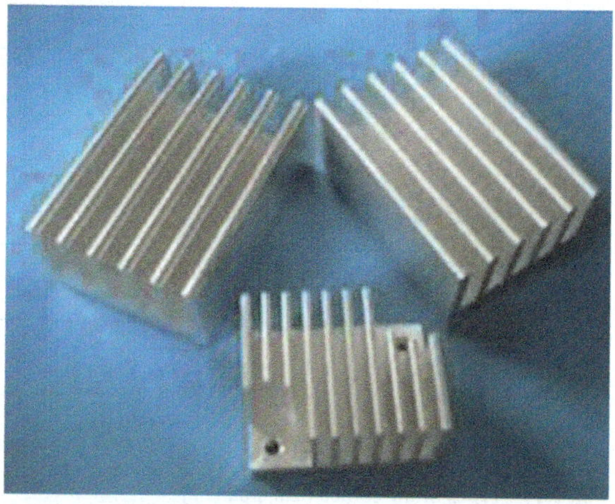

Fig. 1.10 Aluminum heat sink

It showed that different types of channels of the same hydraulic diameter resulted in various frozen coefficients using coolant water in the Silicone Heat Power Cross channel. Difference in shape has affected the properties of heat transfer due to differences in flow dynamics. The tests on the effect of cross-sectional shape were done in non-circular microchannels. The research results showed three ratios of height-to-upper width, the lower total ratio width-to-diameter and the length-to-diameter ratio of the friction factor, and the Nusselt numbers. The numeric flow / heat transfer simulation is performed with water (Re = 100–1000) as a refrigerant on aluminum chain with rhombic, circular, or hexagonal cross section within a hydranged 170–260 μm diameter. The efficiency of the hexagonal channel was higher than the circular and rhombic channel in the case of pressure drop and heat transfer coefficients. However, rhombic channels surpassed other modes in terms of friction, thermal behavior and substratum temperature.

The effect on the heat transfer effectiveness of a silicon sink with the use of laminar coolant (trapezoidal, rectangular and triangular) in the heat sink of silicon (Re = 50–500) is numerically analyzed. The heat transfer effectiveness of the cross-section channels in the triangle was higher when compared to the hydraulic diameters of the rectangular and trapezoidal channels. In terms of the heat transfer performances of the aluminum heat sink of the Reynolds range 100–1000 the effect of the channel cross section (rectangular, triangle or trapezoid), using water as a coolant has been numerically investigated. Then it is possible to use the rectangular shape to increase the total heat transfer. Nevertheless, rectangular, triangular and trapezoidal channels were tested with various hydraulic diameters. The effect on flow and heat transfer on a water cooled copper heat sink was numerically analyzed recently for the cross-section of the rectangular, trapezoidal and triangular channels. The thermal performance of the rectangular channels was found to be lower on the defined hydraulic diameter, number of channels and cross-sectional area compared to trapezoidal and triangular channels. The different types of air-cooled heat sinks are shown in Fig. 1.11.

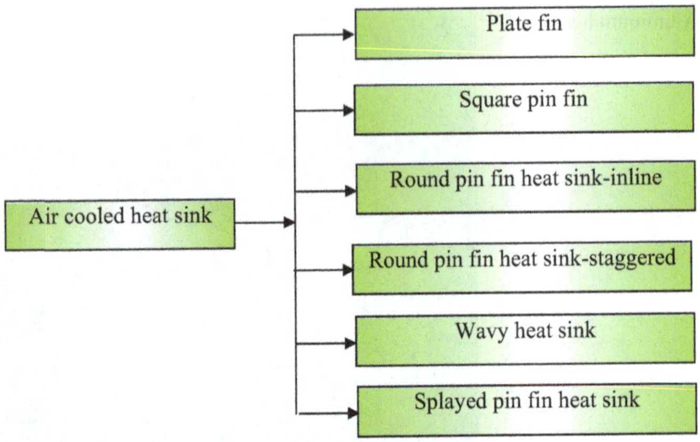

Fig. 1.11 Types of air-cooled heat sink [27]

1.7 Preparation of Nanofluids

One step and two step methods to perform an experimental study or use of different heat transfer application are used to formulation of hybrid nanofluids. The techniques to create hybrid nanofluids are to disperse individual nanoparticles simultaneously or to disperse nanocomposites into the base fluid. Nanofluid applications in many sectors require the refinement of various properties to reach the desired outcomes. The hybrid nanofluids can be well performed and improved results can be obtained by changing the nano-powder levels of each individual [28].

The preparation of nanoparticles into the base fluid is not just a simple process. The stabilization and proper mixing of nanofluids with homogenously dispersed nanoparticles is important in certain environmental conditions. Nanofluid preparation methods are various and can be divided into two primary groups by the number of preparation steps (Fig. 1.12).

1.7.1 Single Step Method

Nanoparticles are prepared and suspended in the base fluid by single steps and processes with conducted simultaneously as shown in the Fig. 1.13. The nanoparticles are kept, dried, transported, and dispersed in only one process, thus increasing the dispersion in the base fluid and reducing concentrations.

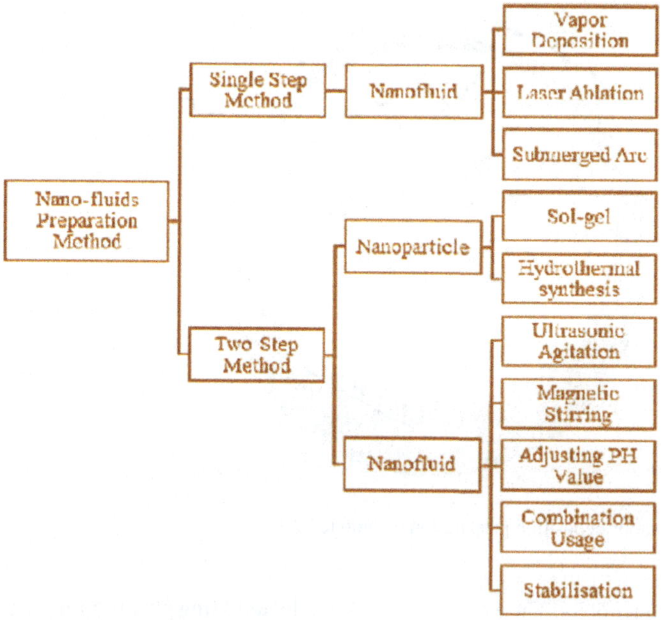

Fig. 1.12 Nanofluids devising methods [29]

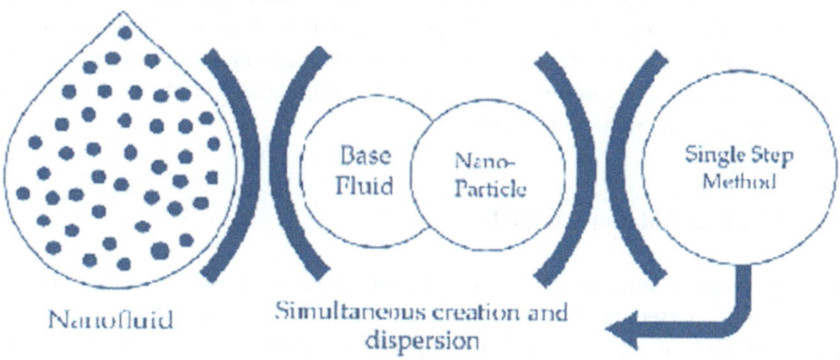

Fig. 1.13 Single step procedure [29]

1.7.1.1 Vapor Deposition

In recent decades, this is the most common one-step procedure. Nanofluid is formed through a thin, swirling water film by means of vapour condensation of the raw material as shown in Fig. 1.14.

A new approach for preparing nanoparticles at high temperatures is a pulsed wire evaporation device (PWE). Four main components are found in the experimental equipment: an EVC, electrical DC with high voltage, a condenser bank and high

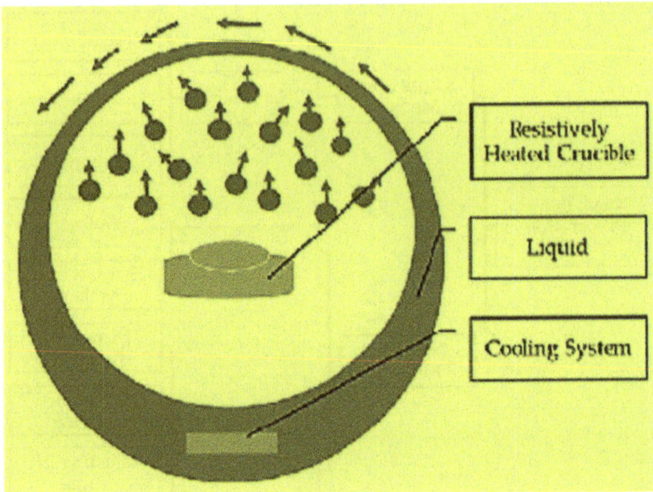

Fig. 1.14 Vapor deposition preparation technique [29]

voltage interrupters. The experiment was conducted using pure titanium nanowires of 5–100 mm diameter and weight. The enhanced thermal conductance has successfully been used to produce ethylene glycol, comprising less than 100 nm nanoparticles. The results show that nanofluid preparation thermal conductivity has been enhanced by increased nanoparticle volumes, but no temperature changes have been seen to have any effect. In preparing nanofluids, but even less common for TiO_2, the vapour repository approach is mainly used as the material in the evaporation chamber had to be evaporated at higher temperatures.

1.7.1.2 Laser Ablation Method

This approach is utilized to disperse the nanoparticles from the surface of the substance immersed in the basic fluid by the highly concentrated laser beam. When the TiO_2 anatine for TiO_2 was dipped, nanofluid TiO_2 was prepared in an from atmospheric nitrogen free conditions as shown in Fig. 1.15.

During the study, the strongest fluid (0.017–0.15 ml/s) in low flow conditions was found in three separate basic fluids (anhydrous ethanol, cyclohexane, water). Due to a sufficient ageing time, the highest laser output potential (100–250 mJ/pulse) and the nano-TiO_2 ethanol colloid were achieved at intensity 414 nm. The laser beam was centered on the titanium plate and the 250 mm quartz focusing lens was passed to meet the experimental needs.

A laser beam was centered on the platform by way of a 250 mm focal lens that vaporizes the plate and spreads nanoparticles in the foundation of the fluid, with total output power of 150 mJ per pulsed. SDS was utilized to create the TiO_2 solution that was cluster-like and stable throughout a week. This is an efficient, cost-effective and much more accurate technique than many conventional methods that involve time and multi-step procedures.

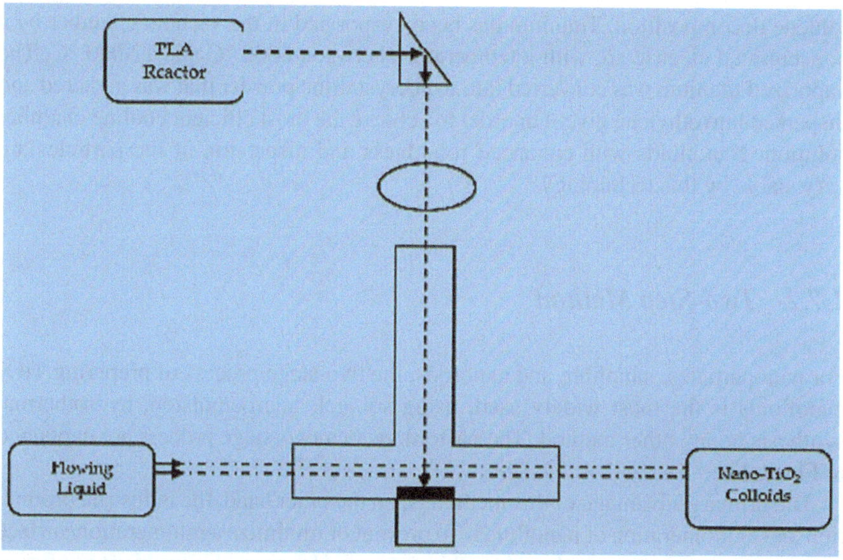

Fig. 1.15 Laser ablation method [29]

1.7.1.3 Submerged Arc Method

The submerged arc method for producing nanofluid is shown in Fig. 1.16. The synthesis method was used for the development of TiO_2 nanofluid using submerged arc nanoparticles; ultraviolet features were investigated from the prepared nano-composite. The TiO_2 nano-composite has been observed with UV radiation during

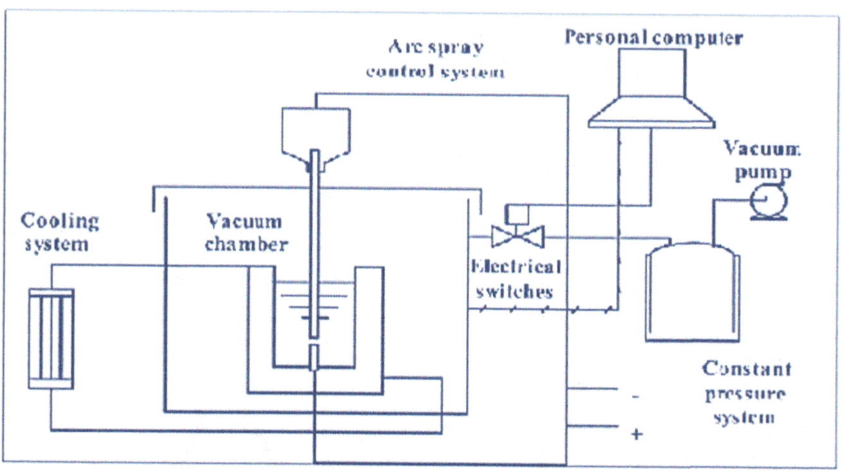

Fig. 1.16 Schematic diagram of submerged arc method [29]

toluene decomposition. Titanium has been evaporated in the vacuum chamber by a concentrated electric arc with a temperature between 6000 °C and 12,000 °C. The vaporized titanium was converted into nanocrystalline powder that was gathered and dispersed into ethylene glycol in order to achieve the most efficient cooling chamber solution. Nanofluids with enhanced roundness and dispersion of the particles are very stable by this technology.

1.7.2 Two-Step Method

For nanoparticles, nanofibre and nanorods, the two-stage process of preparing TiO_2 nanofluids is the most widely used, using sol–gel, microemulsion, hydrothermal synthesis or any other method. The basic steps in a two-stage process are explained in Fig. 1.17.

The primary advantage of this method is, on the other hand, the industrial production and agglomeration of nanofluids. To prevent or minimize agglomeration/surface tension, surfactants are employed for this purpose. However, this problem reinforces the use of a single step technique.

1.8 Characterization of Nanofluids

The characteristics of the collected nanofluids are extremely significant for researchers since experimental research results will be influenced by a variety of nanofluid properties including stability and working cycles as well as the composition of nanoparticles [30].

1. Composition of the Nanofluid
2. Loading of Nanoparticles
3. Impurities in Nanofluid
4. Stability of Nanofluid
5. Thermal Conductivity of Nanofluid

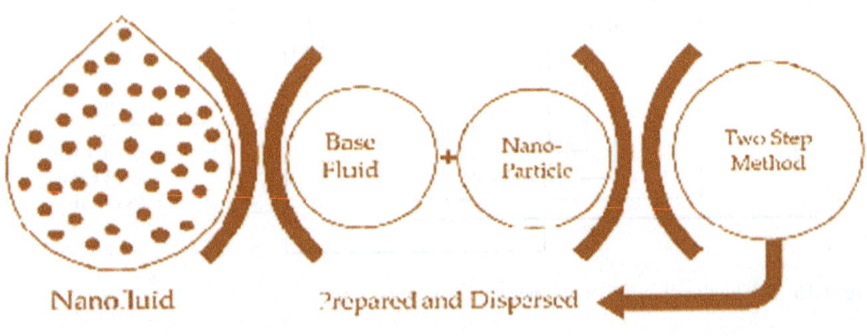

Fig. 1.17 Preparation of nanofluid by two-step method [29]

6. Viscosity of Nanofluid
7. Specific Heat of Nanofluid
8. Specific Gravity of Nanofluid
9. Electrical Conductivity of Nanofluid
10. Magnetic Property of Nanofluid
11. Optical Property of Nanofluid.

1.8.1 Composition of the Nanofluid

Surfactants are categorized based on their polar head in four classes: nonionic surfactants with no head loadings (for example polyethylene oxide, alcohols, etc.), aniotic surfactants with negative head-strips, sulfosuccinate, alkyl sulphate, sulfonate and phosphate, anionic surfactants and long ligated head fatty Acids. The boiling process is mainly affected, in addition to the nanoparticles, by changes in the functional fluid features and changes of the heated surface features due to the deposition of nanoparticles linked to the boiling process. The two mechanisms involve the composition, concentration, scale, shape, surface saturation, texture and geometry, of processing materials (base fluids, nanoparticles). The paths in the base fluid are trapped by the nanofluid. Nanofluids can be seen as an intermolecular power differential as a non-random combination in that the fluid attributes can be represented locally instead of as a whole.

1.8.1.1 Two Liquid Theory

As mentioned in the Fig. 1.18, each molecule is closely circled by other molecules that pertain to each central-local molecule's cell region. In binary combinations

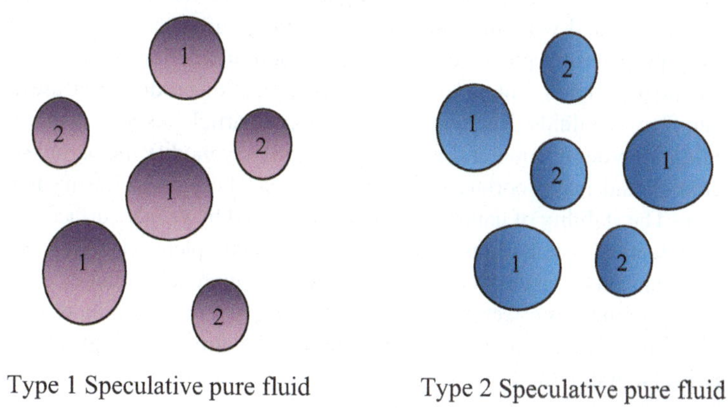

Fig. 1.18 Two types of Speculative pure fluids of binary mixtures

of components 1 and 2 there are two cell groups, Molecule 1 in the middle and Molecule 2 in the middle. The theory can be broadened to include n-component multicomponent mixtures, while pure fluids are not hypothesized so that the J-mix can convey its mathematical properties [31].

1.8.2 Loading of Nanoparticles

Furthermore, after a propanol-based Al_2O_3 nanofluid study, a linear relationship has been identified between density and quantity. This effect has been shown to increase the density to 0.01% and to 0.39% with a weight percentage of carbohydrogen nanotubes, and also for waterborne nanofluid carbon nanotubes. This trend is also applicable to the non-metallic nanofluids, which are proved to be increasing by several studies on carbon nanotube rheology.

Various investigations of the effect of TiO_2 particle loading on the nanofluid viscosity have shown that TiO_2 increases the viscosity with the increase in particles loading. However, some experiments suggest a linear variance in the viscosity ratio, but other observations show that the variation in the viscosity is parabolic. Nanoparticle agglomeration is likely to have a detrimental effect on stability, thermal performance and agglomeration and the degree to which they are loaded [32].

1.8.3 Impurities in Nanofluid

Nanofluid stability can be improved without the use of proper superficial functionality to formulate mechanical equipment for nanofluids using the one-step process. This process, i.e. residual reactants are generally left in nanofluids due to incomplete reactions or stability, is the problem with impurities. The effects of nanoparticles are difficult to understand without reducing this impurity influence. Nanoparticles are intrinsically prepared by processes involving reduction reactions or ion exchange. Moreover, the base fluids contain additional ions or reaction products that are difficult to separate from the fluids. Nanoparticles also apply particle dispersion additives to the base fluid in order to counter this trend. This can may modify the particle surface characteristics and inappropriate impurity levels can be present during nanofluid preparation. The stability of nanofluids can be improved by proper surface functions without mechanical installations. However, by the incomplete reaction or stability of residual reactants in nanofluids, the lower-leaved approach is under privileged. Without eliminating this effect of impurity, the effect of a nanoparticle is difficult to explain. For high temperature applications, the functionality of the high-temperature stabilizer is also important [33].

1.8.4 Stability of Nanofluid

Different forces such as Van der Waal attractive force, gravitational strength, enhanced strength and repulsion are subjected to Sediment and destabilization sensitive liquids with hanging particles. In a variety of applications, the Van der Waal functionality and efficiency of nanofluids is based very much on their stability. The nanofluid stability is caused by dielectric base fluid constant, zeta potentials, pH, particulate matter, and shape, particulate concentration.

For nanofluid stability (Fig. 1.19) large aggregates which lead to the sedimentation of the nanoparticles in the base fluid and nanoparticles should not be produced when suspended (water, oil, alcohol etc.). The theory of Derjaguin, Landau, Vervey and Overbeek (DLVO) states that particles are aggregated and inevitably lead to sedimentation when the attracting potential of Van der Waal is larger than the repellent electrostatic potential.

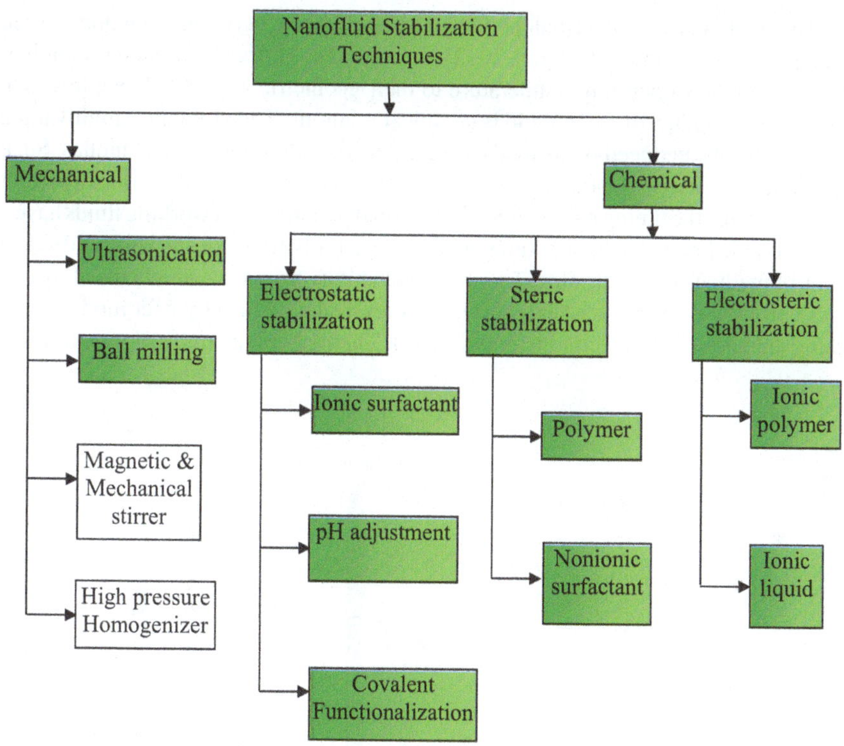

Fig. 1.19 Classification of various nanofluid stabilization techniques [34]

1.8.5 Thermal Conductivity of Nanofluid

The heat transfer characteristics of hybrid nanofluids are defined by the fundamental thermo-physical properties. Thermal enhancement of the nanoparticles is tied to the convection of the base fluid. Good conductivity contributes to the thermal instruments' high performance [35]. There has been a linear relation between volume concentration and thermal conductivity change, represented by Eq. (1.1).

$$k_r = \frac{k_{nf} - k_{bf}}{k_{bf}} = 6.06\phi + 2.72 \tag{1.1}$$

K_r = ratio of thermal conductivity of nanofluid to base fluid
K_{nf} = thermal conductivity of nanofluid
K_{bf} = thermal conductivity of base fluid
ϕ = volume fraction.

The bulk material thermal conductivity difference between commonly used nanoparticles is shown in Fig. 1.20. There are many different direction combinations, from their operating temperature to their geometric size and shape, in which the thermal conduct of a nanofluid is not linear relations. The problem is multivariate and can also be deduced by using data-driven artificial intelligence techniques. Such techniques are used to learn feedback and can train an appropriate algorithm to predict how to change the tuning parameters. The thermal parameters of working fluids have a considerable impact on their cooling performance for both active and passive thermal control methods of electronics. Thermal conductivity primarily affects the ability of these properties to transfer heat. Several studies deal with heat transfer modification using enhanced thermal conductivity of fluid. Thermal conductivity of nanofluids is

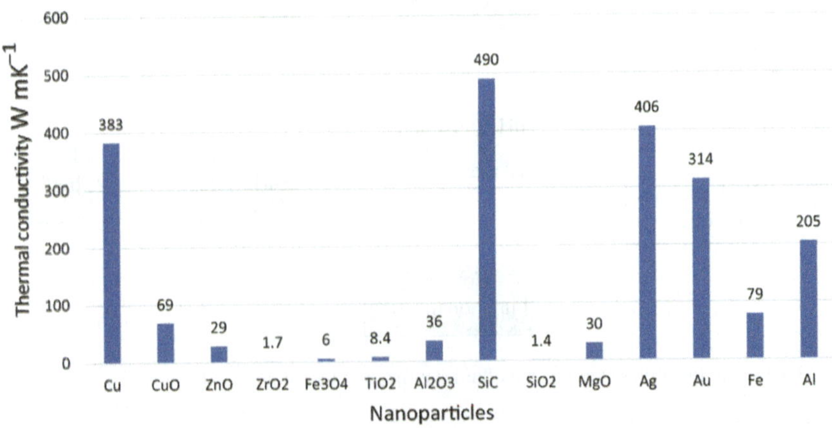

Fig. 1.20 Bulk material thermal conductivity difference between commonly used nanoparticles [36]

subjected to different parameters, such as temperature, solid components, pure, and solid particle fractions.

The use of hybrid nanofluids has been recently the subject of interest and widely explored among different types of nanoparticles due to their low volume level and their high thermal conductivity. The enhanced thermal conductivity of the solid heat source leads to increased heat generation by the energy equation and transferred heat to nanofluid. The solid source of heat for greater thermal conductivity also achieves a better distribution of the temperature [37, 38].

1.8.6 Viscosity of Nanofluid

Nanoparticles showed a low volume fraction of less increased dynamic viscosity, whereas this increase was significant for the higher volume portion. In comparison to spherical nanoparticles, fluids with tube-like nanoparticles have been found to increase viscosity. In addition, the studies showed that they needed to construct their own viscosity models and struggled to create a universal connection instead of using the current models. The forces of Vander Waals cause the formation of clusters among parts that improve the hydrodynamic parts diameter, internal shear stresses, dispersed nanoparticle masses, the particle shape, work and the pH-value of the particles [39]. The fluid viscosity is much higher and impedes Brownian particle motion as well as an optimum concentration value, thereby, reducing thermal conductivity and the heat transfer rate. By increasing the temperature, the viscosity of the fluid decreases and the particle movement is simplified and improved and the heat transfer is increased [3].

The volumetric concentration of nanofluids can be explained by an increase of fluid viscosity (around 1%) which can lead to high wall shearing leading to a thicker limit layer. The rate of heat transfer may therefore be influenced by this phenomenon [40]. The average coefficient of heat loss has been decreased with increased volume. The heat loss coefficient was increased by up to 2.0 wt%. Viscosity declined due to an increase in temperature. Pumping capacity has been improved with the rise in volume. Concentration due to the sudden increase in fluid viscosity was reasonable but because of a high increase in energy efficiency (around 32.23%). At higher concentrations, the efficiency continues to decline as the span between the nanoparticles reduces and in turn causes particles to agglomerate and it results in reduced movement of nanoparticles in Brownian motion [41].

1.8.7 Specific Heat of Nanofluid

Special heat essentially implies the thermal storage potential of the device but can be measured as well, for example, by dynamic thermal conductivity and divergence. Furthermore, efficiency in heat transfer is key to the convective coefficient of thermal transfer and the amount of nanofluids fluid in Nusselt number. The thermal potential of the base fluid, the particulate volume fragment and temperature depend on the thermal strength of the nanoparticles [41]. The special heat capacity of nanofluids is dependant and is determined by the relationship between density and specific heat capacity of the basic fluid is given as,

$$C_{Pnf} = \frac{\varphi \rho_{np} C_{Pnp} + (1 - \varphi) \rho_{bf} C_{Pbf}}{\rho_{nf}} \tag{1.2}$$

Ø = Volume fraction
ρ_{np} = Density of nanoparticle
C_{Pnp} = Specific heat capacity of nanoparticle
ρ_{bf} = Density of base fluid
C_{Pbf} = Specific heat capacity of base fluid
ρ_{nf} = Density of nano fluid.

Specific heat capacity of nanofluids is also dependent on base fluid density and specific heat capacity [42].

Similarly, the heat capacity of the fluid can be increased by nanoparticles. The specific heat of the nanofluid has been increased 50% by doping with exfoliated nanoparticles with a thickness of 100 nm, and a nominal diameter of 20μ with a 0.6 wt% concentration (Table 1.2). Contrary literature studies, however, show that the heat of nanoparticular doping fluids degrades. The related water heat decreases steadily from 0 to 21.7% by as much as 50% when doped with nanoparticles of aluminum Oxide [44]. A well-known correlation for convective heat transfer under turbulent flow condition is

$$\mathrm{Nu} = \mathrm{CRe}^{a}\mathrm{Pr}^{b} \tag{1.3}$$

The Nusselt number expresses in the above equation heat transfer coefficient h for a thermal conductivity fluid k in a pipe diameter d. The fluid flow Reynold numbers varies with the density and dynamic viscosity of average velocity. The number of Prandtl shall be determined by the specific heat. All the thermophysical properties, including the Cp, are based on the temperature and volumetrical concentration of nanofluids.

Table 1.2 Properties of nanoparticles [43]

Nanoparticle	Thermal conductivity (W/mK)	Density (kg/m^3)	Specific heat capacity (J/kgK)
Al_2O_3	36	3920	773
CuO	69	6350	535
TiO_2	8.9	4250	686

1.8.8 Specific Gravity of Nanofluid

In Arctic and sub-Arctic regions worldwide, vast quantities of energy are used to heat industrial and residential structures, such as Alaska, Canada, the regions of Northern Europe and Russia. The heat transfer fluid temperature is below zero in those areas. For calculating the density of nanofluids, a particular bottle method is used. The bottle was filled with water and weighed at different temperatures for calibration. After calibration, the gravity bottle is loaded with nanofluid samples and is weighed again. The real seriousness of the nanofluid is calculated according to the net weight of the deionized water and its net weight. The fraction of nanoparticles experimentally increases, as does the density. Nanofluid density is higher than base fluids, and the density of nanofluid continues to decline as the temperature increases [45].

1.8.9 Electrical Conductivity of Nanofluid

Improving nanoparticle contact via increased concentration and increased nanoparticles conductivity. It indicates that the conductivity of electrical material increases, depending on the types of nanoparticles. As the nanoparticle types or properties of the base fluids were altered (with addition of surfactants) the electrical conductivity was changed entirely. In addition to rheological and thermal characteristics, the electrical properties of nanofluids in particular are studied. In this instance several nanofluid forms, including different types of nanoparticles, a concentration of volume and base fluids have been studied.

There has been evidence that the nonlinear reliance on electric conductivity differs from the reported linear reliance on volume of nanoparticles in ZnO-isolated oil and nanofluid aluminum-oxide water. Nanofluid electricity is rising at temperatures alongside the simple fluid. In addition, an increase in nanofluid volume levels was made in electrical conductivity, close to thermal conductivity. The ability of charge particles, which can carry loads in their respective electrodes, provided by the fluid suspension is a nanofluid electrical conductance. Knowledge of the nanofluid conductivity is important as electricity impacts on the flow and stability of nanofluids [46].

1.8.10 Magnetic Property of Nanofluid

Magnetic nanofluids (also known as ferrofluids or magnetic fluids) are a specific category of intelligent nanomaterials, in particular magnetically controllable nanofluids. These kinds of nano fluids, including Fe_3O_3, γ-Fe_2O_3, $CoFe_2O_4$, Co, Fe or Fe–C, are magnetic nanoparticles that are stably distributed in a carrier liquid. Magnetization curves can be used extensively to manipulate magnetology and magnetological theory to study both particulate and agglomerate interactions. For example, the initial susceptibility is determined by the diameter and dimension of particles, particle interactions, premade aggregates during preparation, or zero or magnetic induced field aggregates development.

The magnetic nanofluids consist of suspension of maghemite, magnetite, cobalt and nozzle in non-magnetic fuel carriers of ferromagnetic nanoparticles. These particles increase the thermal conductivity of nanofluids not just because of their magnetic nature. Ultrasonic and ultrasonic particles can flow into the microchannel without being blocked, and heat transfer equipment can be reduced and heat transfer capacity enhanced simultaneously [47]. The various physical and chemical processes synthesize various nano-fluid types. Nanofluid preparation depends on the agglomeration and various other particle features. Two methods for the preparation of nanofluids are typically available, namely. single step and two step process method. The one-step technique is a method for the combination of nanoparticles and nanofluid synthesis. This ensures that nanoparticles are not concurrently dried, transported, or stored by the production and spread of particles in the base fluid. A two-step process is mainly used in the treatment of nanofluids. The as-prepared nanopowder is then dispersed by ultrasound. The electronic density system has been developed to take into account many aspects of the typical changes in a heat stream, called a heat transmitter. The working fluids (WFs) used to enhance thermal efficiency using nanofluids with outstanding thermal conductivity have been the traditional heat pipe [48].

1.8.11 Optical Property of Nanofluid

In recent years, the mixing of base fluid and nanoparticles has shown that the optical and photothermal conversion properties of transparent fluids which act in nanofluid formation significantly increases. Nanoparticles with selected and adjusted plasmonic and thermal optical property can be developed using the selection of materials, material structure and geometric properties (sizes) including searches for different, better plasma nanoparticles and achieving the "ideal" optical features. Particulate dimensions less effect nanofluid optical characteristics while the extinguishing factor is linearly proportionate to the volume. Although in comparison with absorption the effect of nano article size is small on extinct nanofluids coefficient, the size should be restricted to less than 20 nm to reduce the effects of Rayleigh. Low volume fractions can be employed to avoid difficulties as soon as the optical qualities have been achieved such blocking nanoparticles and unstable suspension [49].

References

1. Chakraverty S, Biswal U (2020) Modeling and simulation of nanofluid flow problems. Synth Lect Mech Eng 5(1):1–89

2. Ahmad S, Nadeem S, Ullah N (2020) Entropy generation and temperature-dependent viscosity in the study of SWCNT–MWCNT hybrid nanofluid. Appl Nanosci 1–13
3. Sajid MU, Ali HM (2019) Recent advances in application of nanofluids in heat transfer devices: a critical review. Renew Sustain Energy Rev 103:556–592
4. Said Z, Assad MEH, Hachicha AA, Bellos E, Abdelkareem MA, Alazaizeh DZ, Yousef BA (2019) Enhancing the performance of automotive radiators using nanofluids. Renew Sustain Energy Rev 112:183–194
5. Koca HD, Doganay S, Turgut A, Tavman IH, Saidur R, Mahbubul IM (2018) Effect of particle size on the viscosity of nanofluids: a review. Renew Sustain Energy Rev 82:1664–1674
6. Sidik NAC, Yazid MNAWM, Mamat R (2017) Recent advancement of nanofluids in engine cooling system. Renew Sustain Energy Rev 75:137–144
7. Dixit T, Ghosh I (2015) Review of micro-and mini-channel heat sinks and heat exchangers for single phase fluids. Renew Sustain Energy Rev 41:1298–1311
8. Adham AM, Mohd-Ghazali N, Ahmad R (2013) Thermal and hydrodynamic analysis of microchannel heat sinks: a review. Renew Sustain Energy Rev 21:614–622
9. Bahiraei M, Heshmatian S, Goodarzi M, Moayedi H (2019) CFD analysis of employing a novel ecofriendly nanofluid in a miniature pin fin heat sink for cooling of electronic components: effect of different configurations. Adv Powder Technol 30(11):2503–2516
10. Salman BH, Mohammed HA, Kherbeet AS (2016) The effect of base fluid type in nanofluids for heat transfer enhancement in microtubes. Appl Mech Mater 818:12–22
11. Rashmi W, Khalid M, Ong SS, Saidur R (2014) Preparation, thermo-physical properties and heat transfer enhancement of nanofluids. Mater Res Express 1(3):032001
12. Amini SM (2019) Preparation of antimicrobial metallic nanoparticles with bioactive compounds. Mater Sci Eng C 103:109809
13. Singh A, Gautam PK, Verma A, Singh V, Shivapriya PM, Shivalkar S, Samanta SK (2020) Green synthesis of metallic nanoparticles as effective alternatives to treat antibiotics resistant bacterial infections: A review. Biotechnology Reports 25:e00427
14. Mody VV, Siwale R, Singh A, Mody HR (2010) Introduction to metallic nanoparticles. J Pharm Bioallied Sci 2(4):282
15. Chavali MS, Nikolova MP (2019) Metal oxide nanoparticles and their applications in nanotechnology. SN Appl Sci 1(6):607
16. Wang S, Gao L (2019) Laser-driven nanomaterials and laser-enabled nanofabrication for industrial applications. In: Industrial applications of nanomaterials. Elsevier, pp 181–203
17. Tian Z, Ge X, Wang Y, Xu J (2019) Nanoparticles and nanocomposites with microfluidic technology. In: Polymer-based multifunctional nanocomposites and their applications. Elsevier, pp 1–33
18. Shojaei TR, Salleh MAM, Tabatabaei M, Mobli H, Aghbashlo M, Rashid SA, Tan T (2019) Applications of nanotechnology and carbon nanoparticles in agriculture. In: Synthesis, technology and applications of carbon nanomaterials. Elsevier, pp 247–277
19. Notarianni M, Liu J, Vernon K, Motta N (2016) Synthesis and applications of carbon nanomaterials for energy generation and storage. Beilstein J Nanotechnol 7(1):149–196
20. Asnawi M, Azhari S, Hamidon MN, Ismail I, Helina I (2018) Synthesis of carbon nanomaterials from rice husk via microwave oven. J Nanomater 2018
21. Wang ZM, Woag A, Salamo G (2009) Lecture notes in nanoscale science and technology. Springer, p 6
22. Al-Waeli AH, Chaichan MT, Kazem HA, Sopian K (2019) Evaluation and analysis of nanofluid and surfactant impact on photovoltaic-thermal systems. Case Stud Thermal Eng 13:100392
23. Yang L, Chen X, Xu M, Du K (2016) Roles of surfactants and particle shape in the enhanced thermal conductivity of TiO_2 nanofluids. AIP Adv 6(9):095104
24. Taha-Tijerina JJ (2018) Thermal transport and challenges on nanofluids performance. Microfluid Nanofluid 1:215–256
25. Fadhil AM, Khalil WH, Al-damook A (2019) The hydraulic-thermal performance of miniature compact heat sinks using SiO_2-water nanofluids. Heat Transfer—Asian Res 48(7):3101–3114
26. https://celsiainc.com/technology/heat-sink/

27. Khattak Z, Ali HM (2019) Air cooled heat sink geometries subjected to forced flow: a critical review. Int J Heat Mass Transfer 130:141–161
28. Babar H, Ali HM (2019) Towards hybrid nanofluids: preparation, thermophysical properties, applications, and challenges. J Mol Liq 281:598–633
29. Ali HM, Babar H, Shah TR, Sajid MU, Qasim MA, Javed S (2018) Preparation techniques of TiO_2 nanofluids and challenges: a review. Appl Sci 8(4):587
30. Sekhar TVR (2015) Characterization of nanofluids-a brief overview. Int J Sci Res Dev 3(07). ISSN (online): 2321-0613
31. Hosseini MS, Mohebbi A, Ghader S (2011) Prediction of thermal conductivity and convective heat transfer coefficient of nanofluids by local composition theory. J Heat Transfer 133(5)
32. Das S, Bandyopadhyay K, Ghosh MM (2018) A study on thermal conductivity and stability of nanofluids containing chemically synthesized nanoparticles for advanced thermal applications. J Mater Eng Perform 27(8):3994–4004
33. Wen D, Lin G, Vafaei S, Zhang K (2009) Review of nanofluids for heat transfer applications. Particuology 7(2):141–150
34. Chakraborty S, Panigrahi PK (2020) Stability of nanofluid: a review. Appl Thermal Eng 115259
35. Sajid MU, Ali HM (2018) Thermal conductivity of hybrid nanofluids: a critical review. Int J Heat Mass Transfer 126:211–234
36. Okonkwo EC, Wole-Osho I, Almanassra IW, Abdullatif YM, Al-Ansari T (2020) An updated review of nanofluids in various heat transfer devices. J Therm Anal Calorim 1–56
37. Tariq R, Hussain Y, Sheikh NA, Afaq K, Ali HM (2020) Regression-based empirical modeling of thermal conductivity of CuO-water nanofluid using data-driven techniques. Int J Thermophys 41(4):1–28
38. Ghaneifar M, Raisi A, Ali HM, Talebizadehsardari P (2020) Mixed convection heat transfer of Al_2O_3 nanofluid in a horizontal channel subjected with two heat sources. J Therm Anal Calorim 1–14
39. Babar H, Sajid MU, Ali HM (2019) Viscosity of hybrid nanofluids: a critical review. Therm Sci 23(3 Part B), 1713–1754
40. Ali HM, Ali H, Liaquat H, Maqsood HTB, Nadir MA (2015) Experimental investigation of convective heat transfer augmentation for car radiator using ZnO–water nanofluids. Energy 84:317–324
41. Wahab A, Hassan A, Qasim MA, Ali HM, Babar H, Sajid MU (2019) Solar energy systems–potential of nanofluids. J Mol Liq 289:111049
42. Arshad W, Ali HM (2017) Experimental investigation of heat transfer and pressure drop in a straight minichannel heat sink using TiO_2 nanofluid. Int J Heat Mass Transfer 110:248–256
43. Ali HM, Arshad W (2015) Thermal performance investigation of staggered and inline pin fin heat sinks using water based rutile and anatase TiO_2 nanofluids. Energy Convers Manage 106:793–803
44. Shin D, Banerjee D (2011) Enhanced specific heat of silica nanofluid. J Heat Transfer 133(2)
45. Sekhar TVR. Characterization of nanofluids-a brief overview
46. Ezekwem C, Dare A (2020) Thermal and electrical conductivity of silicon carbide nanofluids. Energy Sources Part A 1–19
47. Agnihotri P, Lad VN (2020) Magnetic nanofluid: synthesis and characterization. Chem Papers 1–12
48. Chiang YC, Chieh JJ, Ho CC (2012) The magnetic-nanofluid heat pipe with superior thermal properties through magnetic enhancement. Nanoscale Res Lett 7(1):322
49. Said Z, Sajid MH, Saidur R, Mahdiraji GA, Rahim NA (2015) Evaluating the optical properties of TiO_2 nanofluid for a direct absorption solar collector. Numeric Heat Transfer Part A Appl 67(9):1010–1027

Chapter 2
Flow Characteristics of Nanofluids in Heat Sinks

Abstract Heat transfer in fluid flow processes can be divided into two types based on the mechanisms namely, between two fluids and the fluids separated by the channel. Nusselt number, Reynolds number, and friction factor are used to measure the flow properties. According to distinct relations, the heat transfer and Nusselt number are discussed in the chapter. The average Nusselt number and the relationship between Reynolds number, friction factor, and average heat transfer coefficient are furnished in detail.

Nomenclature

χ^*	Thermal length, dimensionless
Nu	Nusselt number
χe	Dryness of the refrigerant
D	Equivalent diameter of channel/m
Re	Reynolds number
Pr	Prandtl number
f	Friction factor, Dimensionless
α	Thermal diffusivity/$m^2 \cdot s^{-1} \cdot K^{-1}$
ε	Porosity
ks	Thermal conductivity/$W \cdot m^{-1} \cdot K^{-1}$
θ_m	Dimensionless temperature
K_{st}	Stagnant thermal
U	Liquid superficial velocities/ms^{-1}
g	Vapor
P	Pressure/Pa
n_f	Nanofluid
h	Heat transfer coefficient
L	Length of the microchannel
ch	Channel
T	Temperature

S. Harikrishnan et al., *Thermal Performance of Nanofluids in Miniature Heat Sinks with Conduits*, Fluid Mechanics and Its Applications 131,
https://doi.org/10.1007/978-981-16-7845-5_2

q	Heat flux
G	Mass flux
ρ	Density
av	Average
Gz	Graetz number
Ø	Nanoparticle volume fraction
WLR	Wave-length ratio
WAR	Wave-amplitude ratio
bf	Base fluid

Contents
This chapter provides information about flow patterns of nanofluids in various conduits, extended surfaces and twisted inserts.

2.1 Introduction

Over the period of enduring years, the heat generation on the electronic parts is prominent. In light illumination of the electronic applications, needs the quick suitable cooling structure to remove heat dissipation from the obliged structure of geometry. To achieve these necessities, the compact heat exchanger device and enhanced heat expulsion fluids is required [1–4]. Accordingly, miniature of the heat transfer device has a pair of trouble. To overcome certain obstruction, the various channels of heat transfer contrivance were utilized to redesign the performance. The competence of heat transfer devices can be improved by using nanofluids. The nanofluids are having anticipated heat removal properties. The dispersion of nanoparticles in base fluids enhances the thermal conductivity of the base fluids [5]. This was accomplished because thermophysical properties including thermal conductivity and viscosity were modified, although special mechanisms are now being evaluated and are not widely understood. Several scientists have examined the explanations underlying better nanofluid heat transfer [6, 7].

A two step method of preparation that reflects the impacts of Brownian motion development and thermophoresis was employed in the view of nanofluid flow in a parallel microchannel plate. Their impacts were shown through a moderation of the importance of the Brown's coefficients of thermophoretic diffusion (N_{BT}). The Brownian motion reduces the concentration of nanoparticles by moving the particles to a lesser obsessive zone, which makes the unification even less uniform [8].

Self diffusion plays a non-irrelevant role in microchannel particle rendezvous by its high reliability of the Peclet number. For the frequently low assessment of Peclet number, substantial extended assortments were obtained essentially under the prohibitive influence of self-distribution. The dimensionless group denoted by N_{BT} (thermophoretic effect) addresses on the movement of particles. It was combated that thermophoresis diffusion clearly depends on tube diameter. The current microchannel results undoubtedly indicate that the thermophoresis activated impacts from the thermal tube wall are relatively restricted and are not significant at all for the heat transfer of 5.0 kW/m^2 [9–12].

The nanofluid is hydrodynamically and thermally presented to the entrance area of the laminar zone. Nanofluids convective heat transfer is modified in the laminar flow and the pressure drop is restricted. Both heat transfer and pressure changes are related to the concentration of nanoparticles volume. For better heat transfer in engineering applications, nanofluids should be used in either laminar or fully developed turbulent flow with Reynolds numbers high enough to encourage turbulent flow (Re) [13, 14]. In case of laminar flow, thermal conductivity (k) is added to the concurrent heat transfer coefficient with an equal development. An indistinguishable divide of heat flow can be replaced more by small diameters [15–20].

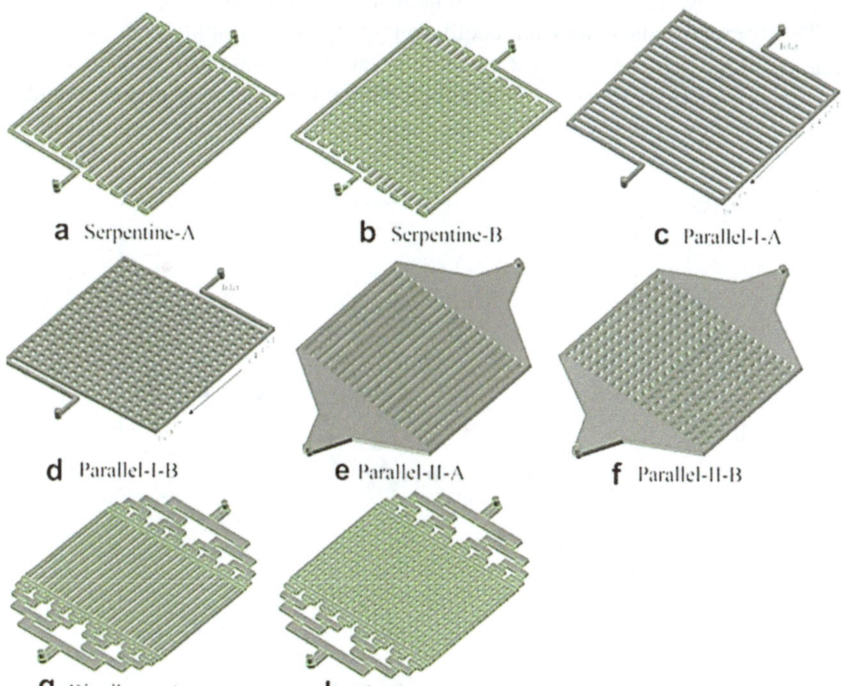

Fig. 2.1 Eight flow distributing layouts [21]

The eight distinct designs combining parallel, bifurcation and serpentine flux for a heat sink based on a microchannel (Fig. 2.1). For information regarding flow performance on channels and pumping energy needed in the same way as surface temperature uniformity. These parameters are useful for estimating these heat exchangers' performance. It was discovered that the most complex styles of carter/bifurcation provide the best pressure factor and surface temperature differential for heat sink applications, making them ideal for liquid-based cooling systems [22].

For measuring the disclosure of heat sinks a division of dimensional variables such as Reynolds number (Re), Prandtl number (Pr) and Nusselt number (Nu) is crucial. In other hand, variety of flow condition can be investigated by using dimensionless variables. The Nusselt number (Nu) of DIW (Deionized Water), 0.1 VF of MgO-DIW, Al_2O_3-DIW and CuO-DIW nanofluids variations is changed with the effect of Reynolds number are showed up in Fig. 2.2. It is clear that, Nusselt number enhanced with the increase of Reynolds number for DIW and all preferred nanofluids. Figure 2.2 shows that Nusselt number for using MgO-DIW and CuO-DIW are slash along that DIW in selected volume flow rate even the assortment between them were originate to be squat. While using Al_2O_3-DIW nanofluids, it shows higher Nusselt number when situate out from DIW in selected volume flow rates. These occurrences are a direct result of higher heat conductivity and viscosity. According to the Reynolds number, the flow was perceived as both laminar and turbulent.

The convective heat transfer coefficient of nanofluids spilling in hexagonal chamber h_{nfHT} were resolved for contrasting the hexagonal chamber side volume flow rate range from 20 to 50 lph and kept volume flow rate in miniature chamber side volume flow speed of 30 lph (Litres per hour). The combinations of h_{nfHT} of the

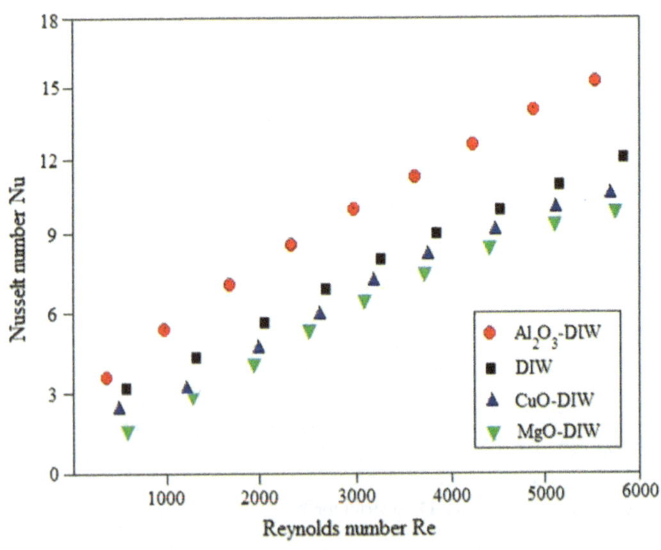

Fig. 2.2 Nusselt numbers versus Reynolds numbers for a constant 30 lph mini passage fluid flow [23]

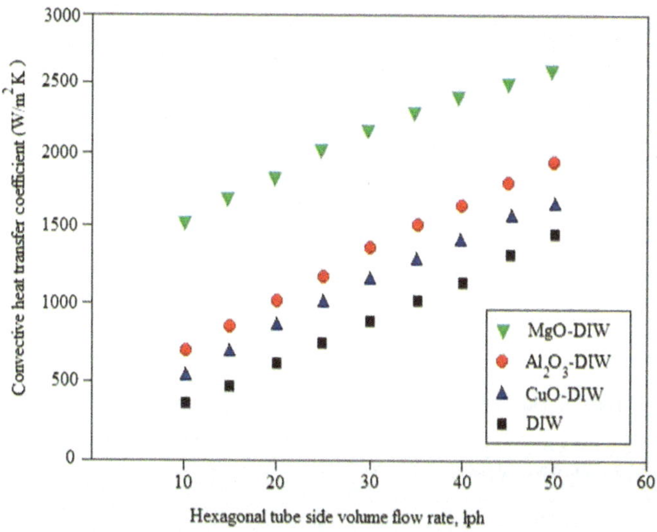

Fig. 2.3 Convective coefficients of heat transfer in various hexagonal tube side fluid volume rates and a 30 lph constant mini passage fluid flow [23]

0.1 VF of MgO-DIW, Al_2O_3-DIW and CuO-DIW nanofluids and DIW for 20 lph to 50 lph of hexagonal tube side flow with consistent 30 lph of miniature chamber side passage are showed up in Fig. 2.3. It could be found in the Fig. 2.3, the heat transfer co-efficient of nanofluid (h_{nfHT}) of chosen nanofluids and DIW is enlarged with the enlargement of volume, flow speed of hexagonal pipe side pour. The h_{nfHT} is found to be higher at volume flow rate of hexagonal pipe side (50 lph of hexagonal chamber side and 30 lph of more modest than normal passage side), in any case, the higher redesign of h_{nfHT} of chosen nanofluids when deviated from DIW was found at 20 lph of hexagonal chamber side and 30 lph of miniature chamber area side flow.

The heat transfer and fluid flow properties of nanofluid and the Nusselt Number (Nu) and friction factor basis for heat transfer have been studied by different researchers. Many other investigators have studied other relationships and they are also displayed in Table 2.1.

2.2 Flow Characteristics

A heat exchanger is characterized by two major phenomena: fluid flow in channels and heat transfer between fluids and channel separation. The thermal transport of heat sinks depends heavily on the geometrical division. The Nusselt number is more affected than the breadth of the channel by recurrence. The nanofluid flow and heat transfer properties in the Triangular Grooved Micro Channel heat sink (TGMCHS) are mainly the result of heat and water-driven break-point layers of re-development,

Table 2.1 Microchannel heat transfer correlation

S. No.	Investigator	Correlation
1	Naphon et al. [24]	$Nu = 2 + 1.1\mathrm{Re}_p^{0.6}\,\mathrm{Pr}^{1/3}$
2	Kuppusamy et al. [25]	$Nu = 1.0598\left[\frac{28.315+27038(x^\circ)+1783300(x^\circ)^2}{1+3049(x^\circ)+472520(x^\circ)^2+35714(x^\circ)^3}\right]$, $(x^\circ) = \frac{x_e}{DR_eP_r}$ $f = 96(1 - 1.3553\alpha + 1.9467\alpha^2 - 1.7012\alpha^3 + 0.9564\alpha^4 - 0.2537\alpha^5)$
3	Pourmehran et al. [26]	$Nu = -\frac{(1-\epsilon)k_s}{\theta_m k_{st}}$, $\theta_m = \frac{\int_0^1 U\theta_{nf}dy}{\int_0^1 Udy}$, $f = \frac{96P}{\mathrm{Re}_{rf}\left[\frac{\alpha}{1+\alpha}\right]}$
4	Hatami and Ganji [27]	$Nu = h\frac{L_{ch}}{k_{nf}} = -\frac{(1-\epsilon)k_s}{\theta_{nf}k_{nf}}$, $f = h_{av}\left[\frac{T_s-T_{nf}}{\frac{q''L_{ch}}{(1-\epsilon)k_s}}\right]$
5	Sakanova et al. [28]	$Nu = Nu_\infty + \frac{0.14G_z}{1+0.05G_z^{2/3}}$ $Nu_\infty =$ $\left[1 - \frac{2.0421}{\alpha} + \frac{3.0853}{\alpha^2} - \frac{2.4765}{\alpha^3} + \frac{1.0578}{\alpha^4} - \frac{0.1861}{\alpha^5}\right]$ $f = \frac{D\Delta P}{2\rho_{av}U_{av}^2L}$
6	Mohammed et al. [29]	$Nu = 2 + 0.6\mathrm{Re}^{0.5}\,\mathrm{Pr}^{0.333}$, $f = \frac{2D\Delta P}{\rho_{nf}U_{nf}^2L_{ch}}$
7	Khoshvaght and Sahamiyan [30]	$Nu = 0.0045\mathrm{Re}^{1.1105}\,\mathrm{Pr}^{-0.3544}(1 - \phi)^{-52.1997}(WLR)^{-0.4195}(WAR)^{-0.5931}$ $f = 0.0003\mathrm{Re}^{0.1301}\,\mathrm{Pr}^{-0.2701}(1 - \phi)^{-16.5937}(WLR)^{-1.6994}(WAR)^{-0.6506}$
8	Ho et al. [12]	$Nu = 1.86\left(\frac{\mathrm{Re\,Pr}}{L_{ch}/D}\right)^{1/3}\left(\frac{\mu}{\mu_s}\right)^{0.14}$ $f = \frac{2D\Delta P}{\rho_{nf}U_{nf}^2L_{ch}}$
9	Ghazvini and Shokouhmand [31]	$Nu = \frac{\epsilon\alpha}{1-\epsilon}\frac{k_s}{k_{nf}}\sqrt{2B_i}\tanh\left(\sqrt{2B_i}\frac{\epsilon\alpha}{1-\epsilon}\right)$ $f = \frac{96P}{\mathrm{Re}_{nf}}\left(\frac{\alpha}{1+\alpha}\right)$
10	Zhai et al. [32]	$f = 34.81713\mathrm{Re}^{-0.90158}(1 + \phi)$ $Nu = 0.290263(\mathrm{Re\,Pr})^{0.391801}(1 + \phi)^{19.8223}$
11	Naphon and Nakharintr [33]	$Nu = \frac{\left(\frac{Q_{av}}{A}\right)D}{k_{rf}(T_{s,av}-T_{bc})}k_{eff} = \left[\frac{k_{np}+2k_{bf}-2\phi(k_{bf}-k_{np})}{k_{np}+2k_{bf}+\phi(k_{bf}-k_{np})}\right]k_{bf}$ $f = \begin{cases}(1+0.15\mathrm{Re}^{0.678}), & \mathrm{Re} \le 1000\\ (0.0183\mathrm{Re}), & \mathrm{Re} > 1000\end{cases}$
12	Karimipour [34]	$Nu = (26.45)(0.64)^{\phi}(169.93)^{-B}(0.85)^{\mathrm{Pr}}$

the generation of vortices and increased heat transfer surface in the part zone. This whole phenomenon improves inside and outside heat transfer while enhancing the decrease in the pressure [35].

With the extension of the point and importance of the achievement, the thermal prologue of the TGMCHS had grown fundamentally. TGMCHS thermal power of Al_2O_3–H_2O is discovered to be out-maneuvered with water from the basic MCHS (Ø = 0.04, dnp = 25 nm) [25].

The thermal conductivity is taken into account as a portion of the nanofluid concentration particle volume center (Cu and TiO_2). Results indicated that the convective heat transfer factor and the number of Nusselt were proportionately increased by 52 and 49%, at 1.0 vol%. Likewise, heat transfer rates were lowered at greater nanofluid volumes.

However, improvement of the coefficient of convective heat transfer of 27% was cultivated with 1.6% volume of nanofluid in respect to that of the base fluid, which was different. The result shows that the incredible heat conductivity in combination with 1.0% volume increased more nanofluid concentrations [36].

Silicone based cross-flow micro heat exchanger with wafers and multiple high-point channels strengthened by diffusion with medium-layer aluminum. By using water as the working fluid indicated that the flow field is successfully laminated and that the best pressure factor drop at flow rate was 2.47 bar, which was more than 4.5 l/min [37].

The results show that TiO_2/Al_2O_3 hybrid-water nanofluid, nanofluid-water TiO_2 and nanofluid-water Al_2O_3 give preferential convective coefficients of heat transfer over the base water. The hybrid Al_2O_3/TiO_2—water nanofluid diverging from the pure water of 40,000 Reynolds was used to improve the convective heat transfer coefficient by 16.97% [38].

The arrangements of triangular and straight shows the better than rectangular, circular and hexagonal channel heat sink as shown on Fig. 2.4. Due to this reality, a three-way microchannel heat sink is best related to the other four shapes. However, the technique to produce a three-way microchannel heat sink is difficult and extravagant; it's not sensitive to supply such heat sink so that a microchannel heating sink may be set up straight [39].

The flow distribution is confined routinely by pressure loss factor difficulties associated with the main method of fluid, and the low resistance to fluid cordialities is quite common. The homogenous fluid distribution in the channels reduces the difference in pressure loss factor contrast. Therefore, the exit flow and intake channel flow must be designed to keep every flow uniform as its shape ensures the distribution of the flow (Table 2.2).

The aim of the design is to be seen on heating surfaces with a better consistent cooling. The distinguished between A and B thermal sinks is established in the conduits with the ultimate goal that the flow goes directly into the A thermal sink while it enters a 90° transformation into the B thermal sink (Fig. 2.5).

Two key criteria in the measurement of heat sink capacity include thermal resistance (R) and temperature dispersion consistency (θ). Figure 2.6 shows the thermal obstacle for the heat sinks at various levels and velocity. The thermal sink B is

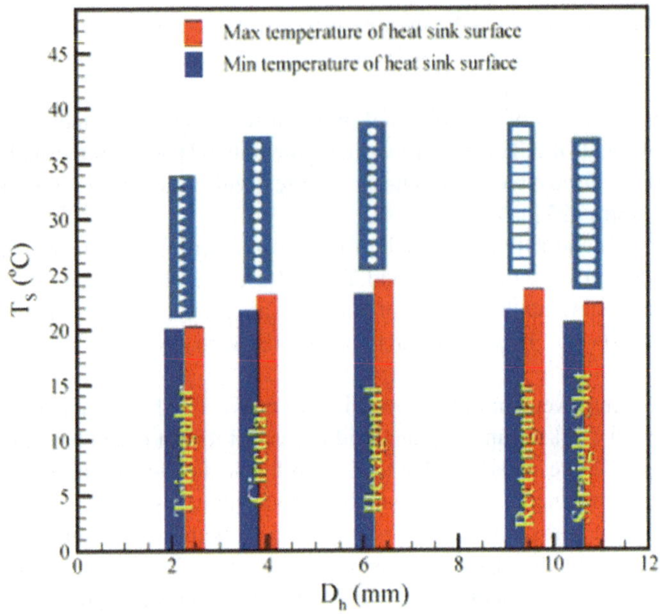

Fig. 2.4 Comparison with the aluminium channel heat sink performance with the varied input pipe for its hydraulic diameter below 60 FCM (cubic feet per minute) [39]

Table 2.2 Design of various flow patterns described by the researchers

Flow patterns design	Flow rate	Thermal/electrical efficiency	Reference
Pan cake	0.0009–0.05 m/s	2%	Naher et al. [40]
Straight	80 ml/min	20.5%	Karami and Rahimi [41]
Helical	80 ml/min	37.67%	Karami and Rahimi [41]
Hexagonal honeycomb	0.11 kg/s	87%	Hussain et al. [42]
Serpentine and hard pipe	–	–	Aste et al. [43]
V-groove type	0.11 kg/s	71%	Yusof et al. & Vaka et al., [44, 45]

definitely correspondingly lower than the thermal sink A and reveals a prominent thermal transfer performed in the thermal sink. It is also noticed that heat resistance diminishes in both heat sinks with an increase in concentration and velocity.

In addition, pumping power (for example, energy consumption) is enormous for assessing the prelude to heat sinks, despite thermal properties. Figure 2.7 displays the pumping power at various obsessions and velocities for two heat sinks. It has seen,

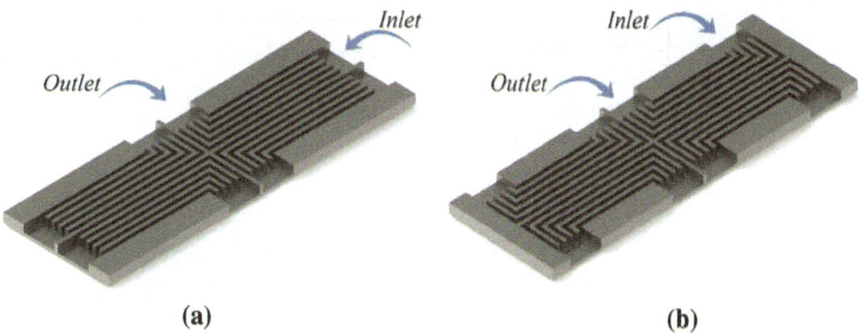

Fig. 2.5 The microchannel heat sinks under study: **a** heat sink A, **b** heat sink B [46]

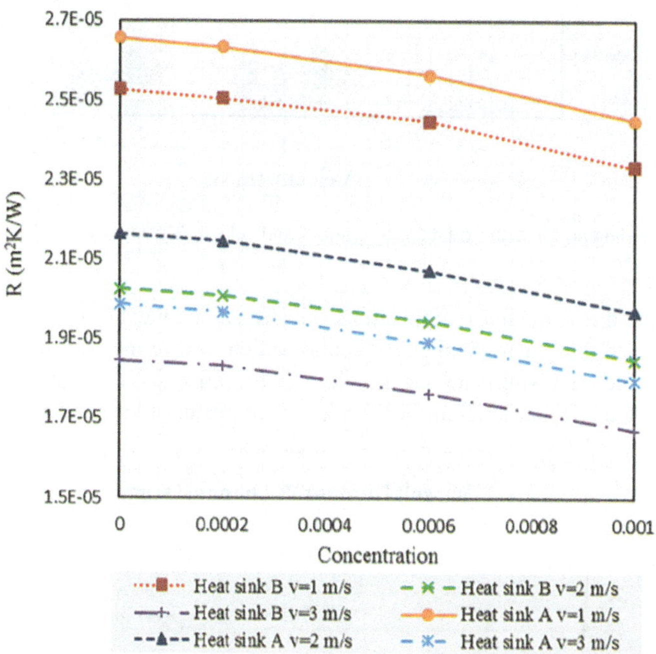

Fig. 2.6 Two heat sinks of heat resistance at varying concentrations and velocity [47]

the increase in velocity greatly increases the pumping power in each heat sink, given the fact that the drop in pressure and the flow rate are increased at higher velocities.

The concentration expansion in both heat sinks improves pumping power. Figure 2.7 shows that heat sinks A are more powerful when pumping, however heat sink B has a course with two abrupt changes, given that the heat sink channels alter with only the isolate path B displayed. The flow pipes in the heat sink B are undoubtedly longer, leading to a better pumping capacity [48].

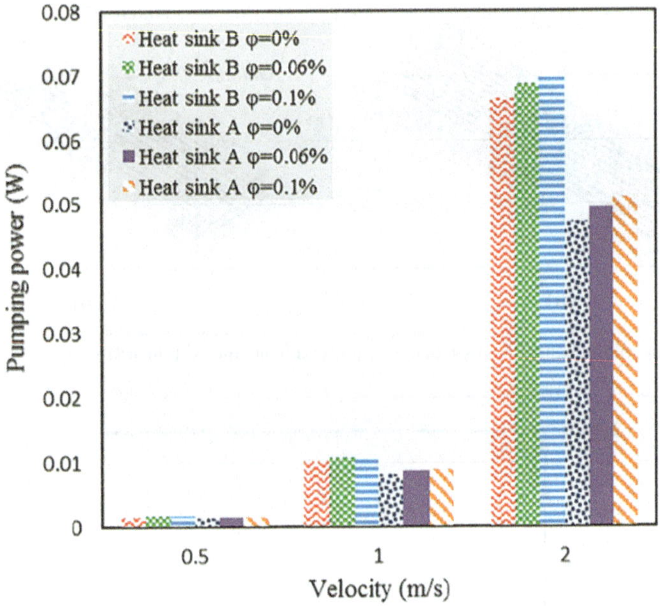

Fig. 2.7 Pumping power with varied concentrations and velocity's for two sinks [47]

To ensure the authenticity of this modernization assessment of three dimensional rectangular straight microchannel, data endorsement was refined for the accuracy of the result. The data is affirmed against Sha and London with the Average Nusselt number evaluated and plotted in Fig. 2.8 as a component of longitudinal distance x.

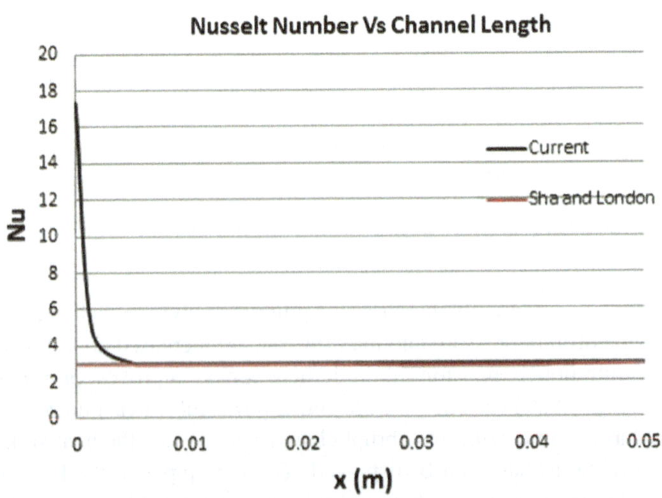

Fig. 2.8 Validation of average Nusselt number along micro channel length [48]

The distinctive Nusselt number from Sha and London [48] have consistent straight line graph while the current evaluation have an ordinary Nusselt number with high value at the part and in the end pushes toward the totally advanced an impetus with extraordinary simultaneousness with Sha and London result as the length increase. The irregularities among Sha and London and this study was a result of the way that, Sha and London applied totally made stream before the working fluid enter the delta however current assessment used uniform inlet speed which require a huge time interval to end up being totally developed taking into account the progression of heat section region at the channel and the assessments of Nusselt number will by and large adjust after totally make province has been developed [49, 50].

Nanofluids can increase heat conductivity and coefficients of heat differ from their basic fluids. The increased viscosity of the fluid caused by the expansion of nanoparticles to a fluid is counteracted by the pumping power. The advantage of more heat transfer compared to the energy discipline having a better commercial sensitivity potential is larger when nano-fluid is used. The pumping power for nanofluids has been viewed as of now for flow in straight chambers (Fig. 2.9) [51].

Evaluation of the cooling performance of an Al_2O_3/water nanofluid microchannel, and the results demonstrated the watercooled nano-fluid heat sink by observing the heat transfer rate and the pressing decrease. The heat transfer rate was quantitatively inspected for a microchannel cooled with Cu / water nanofluids, and it was demonstrated that the solid volume and slanting speed factor in high Reynolds are affected by the heat transfer rate. A dissipating pattern was applied in a circular tube to evaluate the convective laminar flow thermal transfer of nanofluid. A result clearly indicates the essential increase in the thermal transfer of the growth of nanoparticles to the base fluid.

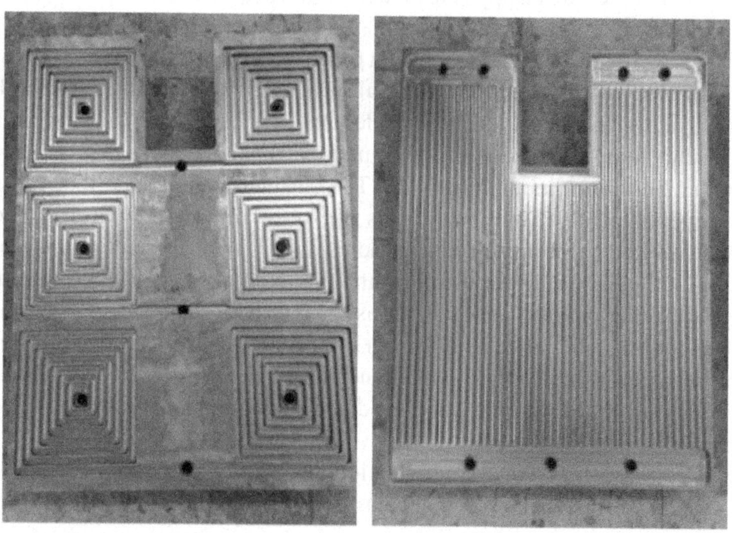

Fig. 2.9 Straight and helical channel refrigeration plate [52]

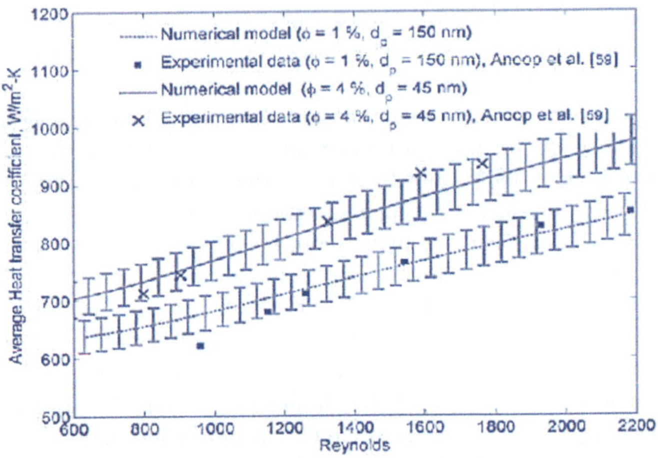

Fig. 2.10 Average coefficient of heat transfer versus numerical and experimental Reynolds number [6]

Lattice Boltzmann Method (LBM) was employed in the manner of a method utilized for simulating flow lines in the rectangular microchannel and examining the effect on development of biological particles of drag, dielectrophoresis and deliciousness. The micro fluidic structure study is in many cases adopted in biomedical and imaginative endeavors because various Micro Electro Mechanical Systems (MEMS) application systems incorporate mass and heat transfer, and molecule polarization. The medical applications of this system helpful and authentic medicine movement, definite and snappy conspicuous confirmation of DNA protein organics, and biological atom action and development predication [53, 54].

In Fig. 2.10, the heat transfer coefficient predicted by this model is generally shown to be consistent with the experimental results in all Reynolds numbers. The problems of thermal management restrict the obstacles to the packaging and mirroring of large-format technology. Liquid cooling with microchannels is part of the selection on exceptional heat sinks in aluminum [6]. The channels can definitely be integrated into a chip, and cooling using nanofluids can be revamped. This work is intended to investigate the improvement of the thermal transfer of the rectangular channel heat sink with laminar flow in view of the pumping power penalty. It includes a description of the progress made in terms of both performance and the technique employed for achieving more modest than standard channel improvements. With the application of graphene-water nanofluid increases the cooling ability and coefficient of the cooler by 72%. In the case of hybrid and aluminum nanofluids, the values are 31.78 and 18.2%.

With a comparative volume centralization of 0.1 vol%, the increase in graphene–water nano-fluid squeezing factor decrease (11.17%) and nano-fluids of the cross

(20.35%) and alumina (33.14%) was found to be lower. Considering the aforementioned, it is advisable for the graphene-water nano-fluid to have a better thermodynamically and performance properties and hence to be reasonable for cooling electronic equipment among the sought nanofluids [55].

Figure 2.11 demonstrates the influence on friction factor of dissolved gas levels after taking account of the variable effects in equation,

$$\frac{f}{f_{cp}} = \left(\frac{\mu_w}{\mu_m}\right)^{0.58} \tag{2.1}$$

f = friction factor
cp = constant properties
μ = viscosity, Pa-s
m = mean
w = wall.

It is shown that with a deteriorated oxygen substance of 8 ppm, the factors of friction are larger, but the results with 1 ppm concern correspond with the predicted laminar flow theory. Thus, it is advised that for testing the friction in microchannels deformed water should be employed. It shows that a test is to reduce the pressure in suitable fluid structures with different gases.

Smaller devices—micro sensors, microscopic valves, micropumps, etc.—are also highly viable as a result of the possible benefits in the use of Micro and Nano Electro Mechanical Systems (MEMS and NEMS). At a level of micro-sizes, the surface impacts increase and change as far as feasible. The slip velocity and temperature spike on the solid–fluid borders constitute the eminent discrepancies between the noticeable ones which are smaller than anticipated. The slip, transient and free molecular flow regime for the gas microflows can be used; in any case, the slip flow regime can be regarded in the most important cases for liquid microfields. Another electromagnetic

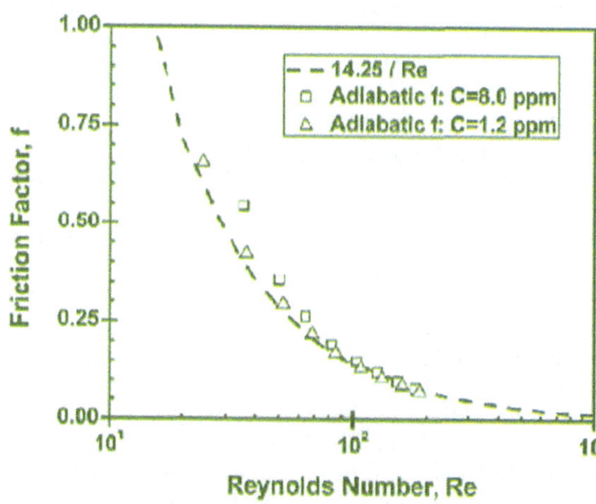

Fig. 2.11 Effect of dissolved gases on microchannel adiabatic factors [56]

flow event is caused by the presence of the electrical double layer (EDL), which is usually implied by a possible flowing effect [57].

References

1. Tuckerman DB, Pease RFW (1981) High-performance heat sinking for VLSI. IEEE Electron Device Lett 2:126–129
2. Kuppusamy NR, Mohammed HA, Lim CW (2013) Numerical investigation of trapezoidal grooved microchannel heat sink using nanofluids. Thermochim Acta 573:39–56
3. Mohammed HA, Bhaskaran G, Shuaib NH, Saidur R (2011) Heat transfer and fluid flow characteristics in microchannels heat exchanger using nanofluids: a review. Renew Sustain Energy Rev 15:1502–1512
4. Salman BH, Mohammed HA, Munisamy KM, Kherbeet AS (2013) Characteristics of heat transfer and fluid flow in microtube and microchannel usingconventional fluids and nanofluids: a review. Renew Sustain Energy Rev 28:848–880
5. Choi SUS, Eastman JA (1995) Enhancing thermal conductivity of fluids with nanoparticles. In: Conference: International mechanical engineering congress and exhibition, San Francisco, CA, United States, 12–17 November. ASME, San Francisco, pp 99–105
6. Mital M (2013) Analytical analysis of heat transfer and pumping power of laminar nanofluid developing flow in microchannels. Appl Therm Eng 50(1):429–436
7. Ahmed M, Eslamian M (2015) Laminar forced convection of a nanofluid in a microchannel: effect of flow inertia and external forces on heat transfer and fluid flow characteristics. Appl Therm Eng 78:326–338
8. Hedayati F, Domairry G (2016) Nanoparticle migration effects on fully developed forced convection of TiO_2–water nanofluid in a parallel plate microchannel. Particuology 24:96–107
9. Lee P-S, Garimella Suresh V, Liu D. Investigation of heat transfer in rectangular microchannels. Int J Heat Mass Transfer 48:1688–1704
10. Chein R, Chuang J (2007) Experimental microchannel heat sink performance studies using nanofluids. Int J Therm Sci 46(1):57–66
11. Jung JY, Oh HS, Kwak HY (2006) Forced convective heat transfer of nanofluids in microchannels. In: ASME International mechanical engineering congress and exposition, vol 47861, pp 327–332
12. Ho CJ, Wei LC, Li ZW (2010) An experimental investigation of forced convective cooling performance of a microchannel heat sink with Al_2O_3/water nanofluid. Appl Therm Eng 30(2–3):96–103
13. Liu D, Yu L (2011) Single-phase thermal transport of nanofluids in a minichannel. J Heat Transfer 133(3)
14. Manay E, Sahin B (2016) The effect of microchannel height on performance of nanofluids. Int J Heat Mass Transfer 95:307–320
15. Chevalier J, Tillement O, Ayela F (2007) Rheological properties of nanofluids flowing through microchannels. Appl Phys Lett 91(23):233103
16. Li J, Kleinstreuer C (2010) Entropy generation analysis for nanofluid flow in microchannels. J Heat Transfer 132(12)
17. Karimipour A, Nezhad AH, D'Orazio A, Esfe MH, Safaei MR, Shirani E (2015) Simulation of copper–water nanofluid in a microchannel in slip flow regime using the lattice Boltzmann method. Eur J Mech B Fluids 49:89–99
18. Malvandi A, Kaffash MH, Ganji DD (2015) Nanoparticles migration effects on magnetohydrodynamic (MHD) laminar mixed convection of alumina/water nanofluid inside microchannels. J Taiwan Inst Chem Eng 52:40–56
19. Zhao G, Jian Y, Li F (2016) Streaming potential and heat transfer of nanofluids in microchannels in the presence of magnetic field. J Magn Magn Mater 407:75–82

20. Alvariño PF, Jabardo JS, Agras JP, Del Valle JG (2015) Self-diffusion assessment in laminar developed flow of nanofluids in microchannels. Int J Therm Sci 98:113–123
21. Ramos-Alvarado B, Li P, Liu H, Hernandez-Guerrero A (2011) CFD study of liquidcooled heat sinks with microchannel flow field configurations for electronics, fuel cells, and concentrated solar cells. Appl Therm Eng 31:2494–2507
22. Siddiqui OK, Zubair SM (2017) Efficient energy utilization through proper design of microchannel heat exchanger manifolds: a comprehensive review. Renew Sustain Energy Rev 74:969–1002
23. Sriharan G, Harikrishnan S, Ali HM (2020) Experimental investigation on the effectiveness of MHTHS using different metal oxide-based nanofluids. J Therm Anal Calorim
24. Naphon P, Wiriyasart S, Arisariyawong T, Nakharintr L (2019) ANN, numerical and experimental analysis on the jet impingement nanofluids flow and heat transfer characteristics in the micro-channel heat sink. Int J Heat Mass Transfer 131:329–340
25. Kuppusamy NR, Mohammed HA, Lim CW (2014) Thermal and hydraulic characteristics of nanofluid in a triangular grooved microchannel heat sink (TGMCHS). Appl Math Comput 246:168–183
26. Pourmehran O, Rahimi-Gorji M, Hatami M, Sahebi SAR, Domairry G (2015) Numerical optimization of microchannel heat sink (MCHS) performance cooled by KKL based nanofluids in saturated porous medium. J Taiwan Inst Chem Eng 55:49–68
27. Hatami M, Ganji DD (2014) Thermal and flow analysis of microchannel heat sink (MCHS) cooled by Cu–water nanofluid using porous media approach and least square method. Energy Convers Manage 78:347–358
28. Sakanova A, Yin S, Zhao J, Wu JM, Leong KC (2014) Optimization and comparison of double-layer and double-side micro-channel heat sinks with nanofluid for power electronics cooling. Appl Therm Eng 65(1–2):124–134
29. Mohammed HA, Gunnasegaran P, Shuaib NH (2010) Heat transfer in rectangular microchannels heat sink using nanofluids. Int Commun Heat Mass Transfer 37(10):1496–1503
30. Khoshvaght-Aliabadi M, Sahamiyan M (2016) Performance of nanofluid flow in corrugated minichannels heat sink (CMCHS). Energy Convers Manage 108:297–308
31. Ghazvini M, Shokouhmand H (2009) Investigation of a nanofluid-cooled microchannel heat sink using fin and porous media approaches. Energy Convers Manage 50(9):2373–2380
32. Zhai YL, Xia GD, Liu XF, Li YF (2015) Heat transfer enhancement of Al_2O_3-H_2O nanofluids flowing through a micro heat sink with complex structure. Int Commun Heat Mass Transfer 66:158–166
33. Naphon P, Nakharintr L (2015) Turbulent two phase approach model for the nanofluids heat transfer analysis flowing through the minichannel heat sinks. Int J Heat Mass Transfer 82:388–395
34. Karimipour A (2015) New correlation for Nusselt number of nanofluid with Ag/Al_2O_3/Cu nanoparticles in a microchannel considering slip velocity and temperature jump by using lattice Boltzmann method. Int J Therm Sci 91:146–156
35. Sajid MU, Ali HM, Sufyan A, Rashid D, Zahid SU, Rehman WU (2019) Experimental investigation of TiO_2–water nanofluid flow and heat transfer inside wavy mini-channel heat sinks. J Therm Anal Calorim 137(4):1279–1294
36. Ganvir RB, Walke PV, Kriplani VM (2017) Heat transfer characteristics in nanofluid—a review. Renew Sustain Energy Rev 75:451–460
37. Mohammed HA, Bhaskaran G, Shuaib NH, Saidur R (2011) Heat transfer and fluid flow characteristics in microchannels heat exchanger using nanofluids: a review. Renew Sustain Energy Rev 15(3):1502–1512
38. Ataei M, Moghanlou FS, Noorzadeh S, Vajdi M, Asl MS (2020) Heat transfer and flow characteristics of hybrid Al_2O_3/TiO_2–water nanofluid in a minichannel heat sink. Heat Mass Transfer 56(9):2757–2767
39. Moradikazerouni A, Afrand M, Alsarraf J, Mahian O, Wongwises S, Tran MD (2019) Comparison of the effect of five different entrance channel shapes of a micro-channel heat sink in forced convection with application to cooling a supercomputer circuit board. Appl Therm Eng 150:1078–1089

40. Nahar A, Hasanuzzaman M, Rahim NA (2017) A three-dimensional comprehensive numerical investigation of different operating parameters on the performance of a photovoltaic thermal system with pancake collector. J Solar Energy Eng 139(3). Article no 031009
41. Karami N, Rahimi M (2014) Heat transfer enhancement in a PV cell using Boehmite nanofluid. Energy Convers Manage 86:275–285
42. Hussain M, Othman YH, Yatim B, Ruslan H, Sopian K, Ibarahim Z (2013) A study of PV/T collector with honeycomb heat exchanger. Proc AIP Conf 1571(1):10–16
43. Aste N, Del Pero C, Leonforte F (2012) Thermal-electrical optimization of the configuration a liquid PVT collector. Energy Proc 1:1–7
44. Othman MYH, Hussain F, Sopian K, Yatim B, Ruslan H (2013) 'Performance study of air-based photovoltaic-thermal (PV/T) collector with different designs of heat exchanger.' Sains Malaysiana 42(9):1319–1325
45. Vaka M, Walvekar R, Rasheed AK, Khalid M, Panchal H (2019) A review: emphasizing the nanofluids use in PV/T systems. IEEE Access 8:58227–58249
46. Bahiraei M, Heshmatian S (2018) Thermal performance and second law characteristics of two new microchannel heat sinks operated with hybrid nanofluid containing graphene–silver nanoparticles. Energy Convers Manage 168:357–370
47. Shah RK, London AL (1978) Laminar flow forced convection in ducts: a source book for compact heat exchanger analytical data, vol 1. Academic Press
48. Abubakar SB, Sidik NC (2015) Numerical prediction of laminar nanofluid flow in rectangular microchannel heat sink. J Adv Res Fluid Mech Therm Sci 7(1):29–38
49. Kang SW, Chen YT, Chang GS (2002) The manufacture and test of (1 1 0) orientated silicon based micro heat exchanger. Tamkang J Sci Eng 5(3):129–136
50. Kang SW, Tseng SC (2007) Analysis of effectiveness and pressure drop in micro crossflow heat exchanger. Appl Therm Eng 27:877–885
51. Routbort JL, Singh D, Timofeeva EV, Yu W, France DM (2011) Pumping power of nanofluids in a flowing system. J Nanopart Res 13(3):931–937
52. Salem MR, Ali RK, Elshazly KM (2017) Experimental investigation of the performance of a hybrid photovoltaic/thermal solar system using aluminium cooling plate with straight and helical channels. Sol Energy 157:147–156
53. Yang YT, Lai FH (2011) Numerical study of flow and heat transfer characteristics of alumina-water nanofluids in a microchannel using the lattice Boltzmann method. Int Commun Heat Mass Transfer 38(5):607–614
54. Ay C, Young CW, Young CF (2012) Application of lattice Boltzmann method to the fluid analysis in a rectangular microchannel. Comput Math Appl 64(5):1065–1083
55. Ahammed N, Asirvatham LG, Wongwises S (2016) Entropy generation analysis of graphene–alumina hybrid nanofluid in multiport minichannel heat exchanger coupled with thermoelectric cooler. Int J Heat Mass Transfer 103:1084–1097
56. Mala GM, Dongqing L, Dale JD (1997) Heat transfer and fluid flow in microchannels. Int J Heat Mass Transfer 40(13):3079–3088
57. Kandlikar SG (2012) History, advances, and challenges in liquid flow and flow boiling heat transfer in microchannels: a critical review. J Heat Transfer 134(3)

Chapter 3
Flat Fin Heat Sink with Forced Convection

Abstract A miniature microchannel heat sink with forced convection is an efficient way to dissipate the heat at a faster rate in electrical and electronic components. The vast majority of semiconductor-based electronic devices generate a considerable amount of waste heat. The conventional microchannel heat sinks are available in different profiles such as trapezoidal, rectangular, and triangular structures. In addition, the design variables and governing equation of the miniature heat sinks are presented herein.

Nomenclature

f	Apparent friction
L	Total length of fin
b	Width of the heat sink gap (m)
H	Height of fin
Δp	Pressure drop
k_c	Coefficient of concentration
ke	Coefficient of expansion
V_{ch}	Channel velocity
A	Heat transfer surface area
ρ_{nf}	Density of nanofluid
ρ_p	Density of particle
$C_{p,nf}$	Specific heat of nanofluid
k_f	Thermal conductivity of fluid
μ_{nf}	Viscosity of nanofluid
ϕ	Particle weight concentration
ρ_f	Density of fluid
k_f	Thermal conductivity of nanofluid
μ_f	Viscosity of fluid
M	Molar mass of the base fluid
N	6.022×10^{23} = Avogadro's number

S. Harikrishnan et al., *Thermal Performance of Nanofluids in Miniature Heat Sinks with Conduits*, Fluid Mechanics and Its Applications 131,
https://doi.org/10.1007/978-981-16-7845-5_3

L	Length of MCHS (m)
f_1	= First objective function
PP	Pumping power
Q	Volumetric flow ratio
V	Average inlet velocity
Ac	Cross sectional area of MCHS

Contents

This chapter reveals the modification of design made in miniature heat sink and fabrication process of various conduit miniature heat sinks.

3.1 Introduction

Each electrical and electronic components of the circuit generates heat while supplying electricity. The heat dissipation capability of these parts is insufficient to dispel the heat generated by the high performance semiconductor systems such as power transistors and optoelectronics such as diodes. The result is a premature failure caused by component heating that can lead to a breakdown of the whole circuit or device.

3.2 Flat Fin Heat Sink with Forced Convection

The heat transfer efficiency of different vertical finned heat sink configurations under natural convection is analyzed [1, 2]. Figure 3.1 depicts the fin configurations used

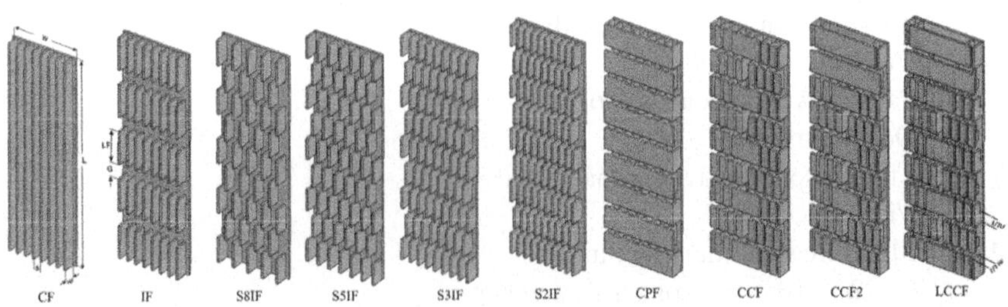

Fig. 3.1 Schematic heat sink configurations [2]

in different specifications. The findings revealed that reducing the space between disrupted fins had no effect on cooling. The heat transfer rate was improved by using capped fins, but they were heavier than continuous fins [3]. Similarly, L-shape and cut-capped fins, on the other hand, are lighter than other fins while offering improved heat transfer.

Six fins were produced in various configurations, including fin numbers and fin spacing. The heat output ranged from 10 to 120 watts. The plate cubic pin fin increased heat transfer by 10 to 41.6% and had lower thermal resistance than a typical plate fin, according to the findings. Additionally, as the number of fins and fin spaces increased, the thermal efficiency improved [4].

As shown in Fig. 3.2, triangular, rectangular, and trapezoidal fins with constant volume were examined [5]. The convection heat transfer was improved by using longitudinal different form perforation with different locations (Fig. 3.3). As the study results showed, the best design, which produced the highest heat transfer rate, was a rectangular perforation with a 0.3 (Hc/H) position.

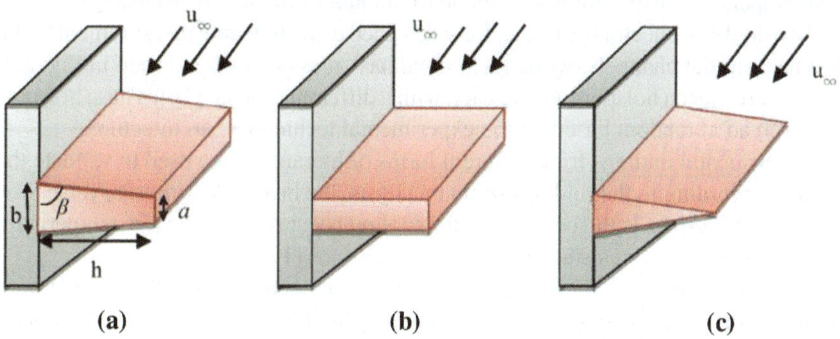

Fig. 3.2 Fins with different geometries: **a** trapezoidal, **b** rectangular and **c** triangular [7]

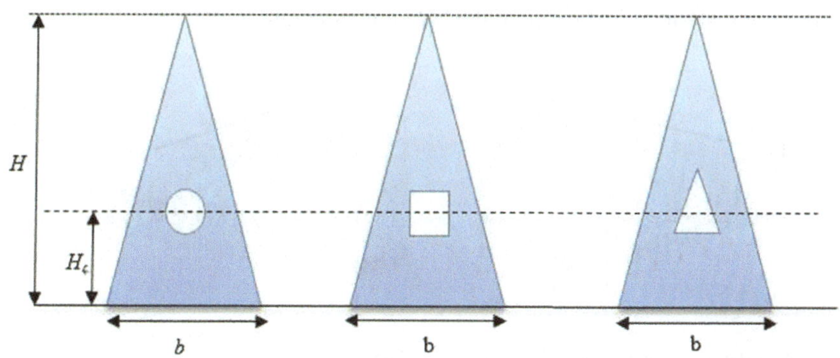

Fig. 3.3 Triangular fin with various shapes of piercing [7]

The heat transfer coefficient, Reynolds number (Re), and Nusselt number (Nu) for perforated fin heat sinks under forced convection were computed [6]. With a constant heat input of 100 W, the Reynolds number was varied from 8000 to 35,000. The fin with circular perforation dissipated more heat than the other configurations, according to the findings [7, 8].

Experiments were performed with various heat flux values ranging from 200 to 360 W and different airflow velocities. The heat transfer coefficient of circular notch fins ranges from 10.34 to 10.55 W/m^2 °C, compared to 10.08 to 10.29 for triangular notch and 9.6 to 9.76 for fins without a notch, according to the findings [9, 10].

With Reynolds numbers ranging from 50 to 500, the Nusselt number, friction factor, and thermal resistance were investigated [11, 12]. The inverse trapezoidal design had the best thermal efficiency and the highest Nusselt number with the lower increase in thermal resistance, according to the findings.

The trapezoidal cross-section configuration had the lowest thermal resistance [13, 14]. Different coolant fluid flow rates and convergence angles were investigated. According to the findings, the highest convergence angle is 4°, which results in the best temperature distribution with the least amount of thermal resistance [6, 15].

Two double-collector heat sinks were designed using the same measurements. The first has parallel channels, while the second has cross channels, as seen in Fig. 3.4. There were eleven hotspots added, each with a different scheme [16]. These hotspots were spread at random based on the experimental techniques' architecture.

Experimental findings from different forms of literature were used to validate the model. According to the findings of that analysis, the heat sink with transverse flow and several hotspots is the best option for cooling because it has the low temperature variance and thermal resistance of the other schemes [17–19].

As shown in Fig. 3.5, five double layer microchannel heat sink (DL-MCHS) were created with triangular, trapezoidal, rectangular, circular, and reentrant shapes. For all of these configurations and the single layer, the thermal resistance, pressure drop, and wall temperature distribution were compared. The rectangular shape was found to be the best choice for the best thermal efficiency among the five DL-MCHS geometries, while the re-entrant shape required the least pumping power [20, 21].

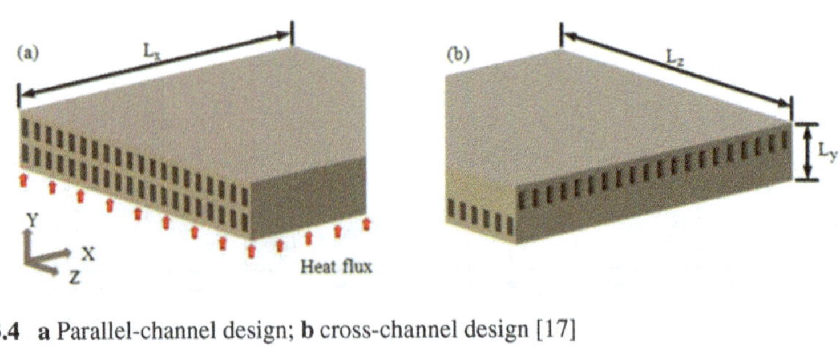

Fig. 3.4 **a** Parallel-channel design; **b** cross-channel design [17]

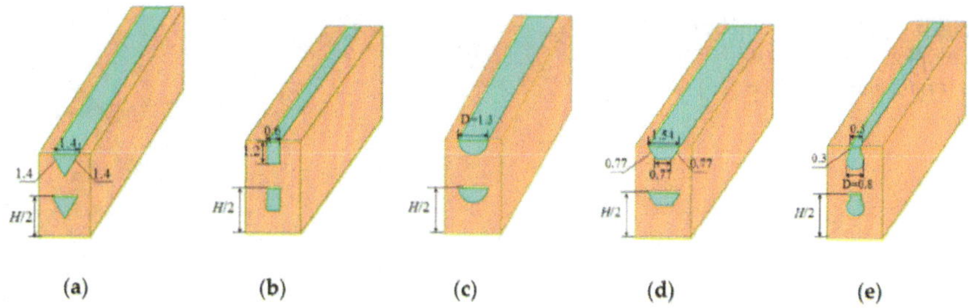

Fig. 3.5 **a** Triangular; **b** rectangular; **c** circular; **d** trapezoidal and **e** re-entrant shaped [22]

3.3 Optimized Heat Sink Topology

Experimental and numerical examinations were done on heat sinks with different airflow rates. It includes thermal resistance and pressure drop comparisons to typical machining designs that use straight plate, radial plate, stepped straight plate, and staggered pin fin configurations.

In laminar flow conditions, various optimized designs with Grashof numbers ranging from 103 to 106 are implemented. This study employs a mechanism that prevents internal cavity creation, which also yields physical effects and observations as well as length-scale effects and forced convection assumptions [22, 23].

The findings were compared to those obtained using the COMSOL program's finite element analysis method and found to be in strong agreement. It was discovered that when Grashof numbers increased, the number of branches on the heat sink also increased. In addition, complicated geometries can greatly improve the cooling performance in comparison to basic fin geometry [24].

The researchers found that the optimized heat sink's thermal resistance is 15% lower than the typical heat sink, making the overall savings 26% (Fig. 3.6). As shown in Fig. 3.7, three aluminum heat sinks were planned, the first of which was an L-shape heat sink, and the others were topology optimized using the robust and non-robust approaches.

The robust method gives the optimized heat sink sharper edges. There are no areas of intermediate densities around these sharp edges. As a result, they are simple to

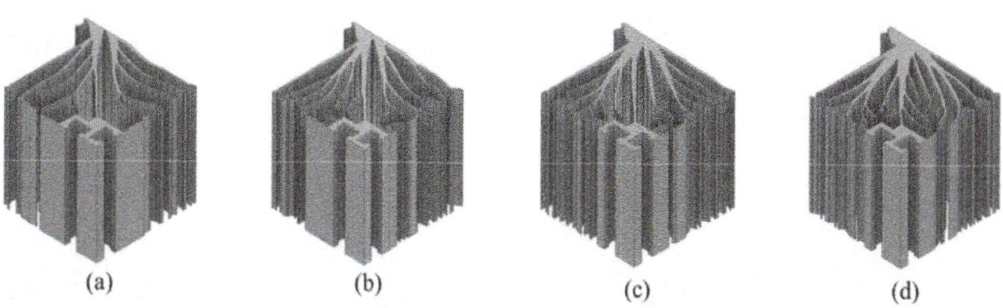

Fig. 3.6 Optimized 3D topology heat sink with volume fraction **a** 0.19, **b** 0.23, **c** 0.27, **d** 0.31 [25]

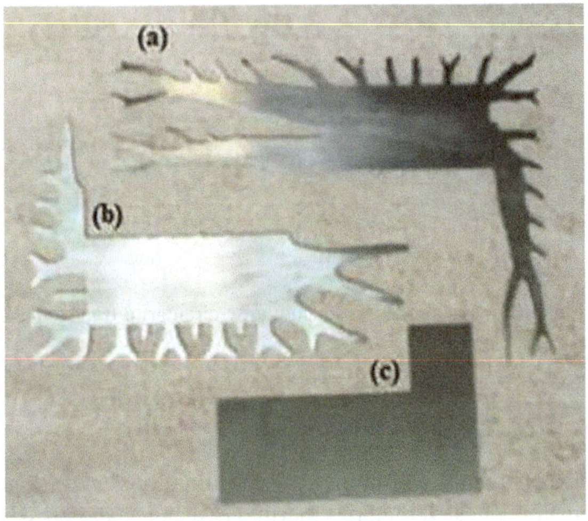

Fig. 3.7 The aluminum heat sink plat: **a** non-robust, **b** robust and **c** L-shape [26]

produce. On conventional straight channels, the optimized MCHS was performed for various velocities. According to the findings, the optimized MCHS saves about 50% of pumping power while maintaining high thermal efficiency [27].

Continuity equation for the nanofluid

$$\nabla.\vec{V} = 0 \tag{3.1}$$

Momentum equation for the nanofluid [28],

$$\rho_{nf}\left(\vec{V}.\nabla\right)\vec{V} = -\nabla P + \mu_{nf}\nabla^2\vec{V} \tag{3.2}$$

Energy equation for the nanofluid

$$\rho_{nf}C_{P,nf}\left(\vec{V}.\nabla\right)T_{nf} = k_{nf.eff}\nabla^2 T_{nf} + \mu_{nf}\left(\frac{\partial V_i}{\partial x_j} + \frac{\partial V_j}{\partial x_i}\right)\left(\frac{\partial V_i}{\partial x_j}\right) \tag{3.3}$$

Energy equation for the solid region

$$k_s\nabla^2 T_s = 0 \tag{3.4}$$

The boundary condition are as follows:

$$V_{in} = u_{in},\, T = T_{in}\ \text{at x} = 0 \tag{3.5}$$

$$\vec{V} = 0,\ Ts = T_{nf},\, k_s\nabla T_s = k_{nf.eff}\cdot\nabla T_{nf} \tag{3.6}$$

at the channel wall-fluid interfaces

$$q_W = -k_s \frac{\partial T_s}{\partial y} \tag{3.7}$$

at the bottom wall of heat sink

$$\nabla T_s = 0 \tag{3.8}$$

at other outside walls of heat sink

$$\mathrm{P} = \mathrm{P_{out}} \quad \mathrm{at\ x} = \mathrm{L_x} \tag{3.9}$$

The following equation has been employed to estimate the effective thermo-physical properties of the nanofluid [29]:

$$\rho_{nf} = \phi\rho_p + (1-\phi)\rho_f \tag{3.10}$$

$$C_{P,nf} = \phi\rho_p C_p + (1-\phi)\rho_f\left(\frac{C_{pf}}{\rho_{nf}}\right) \tag{3.11}$$

$$k_{nf} = k_f\left[\frac{k_p + (n-1)k_f - (n-1)\phi(k_f - k_p)}{k_p + (n-1)k_f + \phi(k_f - k_p)}\right] \tag{3.12}$$

n = the nanoparticle sphericity and is equal to $3/\psi$

$$\mu_{nf} = \mu_f\left[\frac{1}{1 - 34.87\left(\frac{dP}{df}\right)^{-0.3}\phi^{1.03}}\right] \tag{3.13}$$

where the equivalent diameter of the basefluid (d_f) is calculated as:

$$d_f = 0.1\left[\frac{6M}{N\pi\rho_f}\right]^{1/3} \tag{3.14}$$

3.4 Governing Equations

Fluid motion equations are usually complex in nature, necessitating a high level of computing capacity and expertise to achieve better and more reliable prediction performance. As applied to fluids in motion under pressures, this equation is derived from Newton's second law. The Navier–Stokes equation is derived from the stresses caused by the interaction of the diffusion viscous term and pressure [30]:

$$\frac{\partial \rho}{\partial t} + \nabla.(\rho u) = 0 \tag{3.15}$$

$$\frac{\partial u}{\partial t} + (u.\nabla)u = \frac{-1}{\rho}\nabla \rho + F + \frac{\mu}{\rho}\nabla^2 u \tag{3.16}$$

$$\rho\left(\frac{\partial \varepsilon}{\partial t} + u.\nabla\varepsilon\right) - \nabla(k_H \nabla T) + \rho\nabla u = 0 \tag{3.17}$$

where

$$\nabla() = i\frac{\partial}{\partial x} + j\frac{\partial}{\partial y} + k\frac{\partial}{\partial z} \tag{3.18}$$

$$\Delta p = \left(\frac{f_{app}.N(2HL + bL)}{HW} + k_c + k_e\right)\left(\frac{1}{2}\rho V_{ch}^2\right) \tag{3.19}$$

$$\Delta p = \left(4 f_{app} x^+ + k_c + k_e\right)\left(\frac{1}{2}\rho V_{ch}^2\right) \tag{3.20}$$

$$\Delta p = 4\left(f_{app} x^+ + k_c + k_e\right)\left(\frac{1}{2}\rho V_{ch}^2\right) \tag{3.21}$$

$$V_{ch} = V_{ap}\left(1 + \frac{t_f}{b}\right) \tag{3.22}$$

$$V_{ch} = V_{ap}\left(\frac{A_f}{A_{tf}}\right) \tag{3.23}$$

3.5 Design Variables

The channel height (Hc), channel width (Wc), and Reynolds number are the design parameters (Re). The spacings between the micro-channels (Ww) are determined by the symmetry boundary condition and the micro-channel width. The height of the micro-channel from the bottom surface (H_b) is kept constant (Table 3.1).

The objective functions are:

$$N\overline{u} = \frac{1}{L}\int_0^L Nu(z)dz \tag{3.24}$$

Table 3.1 The upper and lower limits of the feature variables [31]

	Wc (mm)	Hc (mm)	Re
Lower bound	0.1	0.4	500
Upper bound	0.3	1	1000
Step size	0.001	0.015	50
Qu and Mudawar [18]	0.231	0.713	139–1672
Wang et al. [32]	0.09–0.4	0.4–1.830	100–1000

$$f_1 = -\frac{N\overline{u}}{N\overline{u}_{ref}} \tag{3.25}$$

$$PP = \Delta PQ = \Delta PVA_c \tag{3.26}$$

$$f_2 = \frac{PP}{PP_{ref}} \tag{3.27}$$

3.6 Straight Channel

The modification in channel size has demonstrated that changes to the microchannel's heat transfer output induce flux distribution differences. The straight microchannel heat sinks configurations are shown in Fig. 3.8.

However, the non-uniform distribution of the flow through parallel heat sinks of the micro-channel, also known as mis distribution of the flow, may lower the efficiency of parallel heat sinks.

Fig. 3.8 Heat sink straight micro-channel [33]

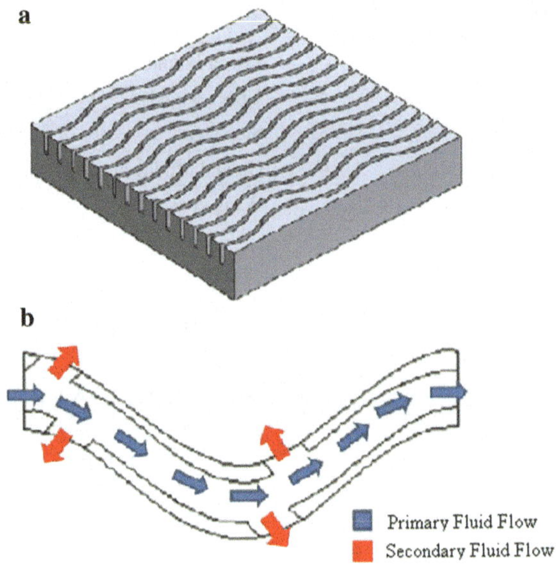

Fig. 3.9 **a** Wavy design, **b** secondary branch [33]

3.7 Wavy Channel

In the 1980s, wavy wall adaptations, re-entry cooling, ablative cross-hatching surfaces and chamber vaporization for the fuel chamber were commercially employed for aircraft. In addition, heat generated by electronic gadgets can be dissipated. An additional microchannel thermal sink design, as shown in Fig. 3.9a. As seen in Fig. 3.9b, a wave design can create Dean's vortices, which help fluid mixes in the channel.

3.8 Pin–Fin

The efficiency of heat evacuation of the pin (Fig. 3.10) Heat sink, for low pumping power, is better than the pin–fin microchannel, and high pumping power vice versa. The heat sink construction can be optimized by minimizing entropy generation. There were no vortices created in the area of cavities when the Reynolds number is less than 79.5. As Reynolds number increase, turbulent flow intensifies and heat transfer improves in the voids.

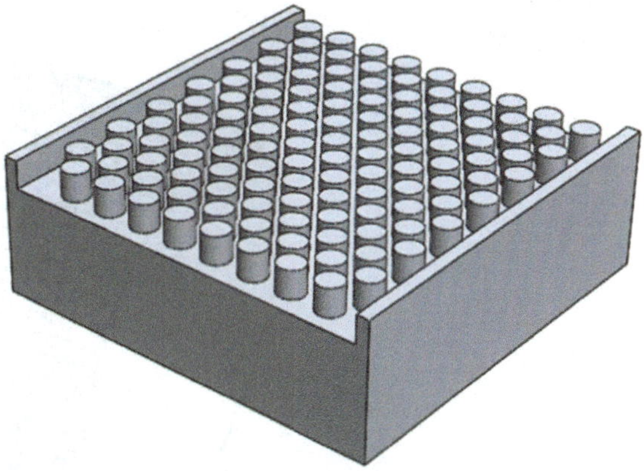

Fig. 3.10 Pin fin micro-channel heat sink [33]

3.9 Ribs

The utilization of cavities and ribs in the microchannel would improve the effectiveness of the heat microchannel compared with individual techniques. The recirculating vortices improved the thermal efficiency of the droplet-based system, while the gutter slowed the heat transfer operation. Furthermore, it was discovered that transporting hot fluid to the droplet's center and moving fresh fluid to the wall can enhance heat transfer in the slug region between droplets. Figure 3.11 illustrates the micro-channel heat sink sinusoidal cavities.

Thermal efficiency in microchannels has become a serious challenge due to the widespread use of microfluid systems in various heat transfer applications. The Navier–Stokes three-dimensional analysis examines the effect of ribbed cables on the effectiveness of a heat sink. In laminar flow, six rib shapes were used with Reynolds numbers ranging from 100 to 500. The measurements of the ribs were set, and the results were obtained using the conventional channel [34].

The triangular rib had the lowest thermal resistance, while the rectangular rib had the highest pressure drop among the different rib configurations. The heat transfer efficiency and fluid flux features of a micro-channel heat sink utilizing both experimental and analytical approaches with complicated onset channels are assessed. The thermal resistance and enhancement amount were measured and compared to

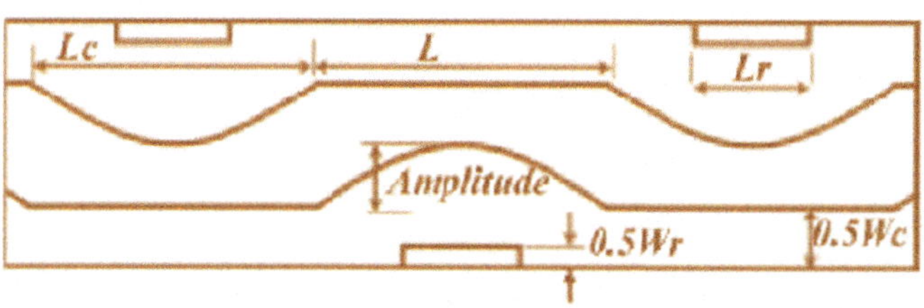

Fig. 3.11 Geometric design of the sinusoidal cavities [33]

Fig. 3.12 Cylindrical oblique fin [34]

a smooth rectangular tube. The measurements of the two channels were identical, and the corrugation max/min widths were 0.2/0.1 mm with a 0.2 mm pitch. Thermal efficiency for the corrugated channel was improved by 1.24% for Reynolds number equal to 611, and thermal resistance was decreased by 18.99% compared to the standard rectangular channel, according to the findings.

3.10 Oblique Fin Channel

The enhancement factor for heat transfer in Reynolds number 300 is decreased to 1.6 compared to a standard heat sink. In research work, the effectiveness of thermal translation of the cylindrical oblique fin (Fig. 3.12) and traditional straight channel was analyzed.

There are several micro-channel designs that support the improvement of heat transfer phenomenon. One design is recommended, using a tiny channel in silicone with staggered, stepped fins, as illustrated in Fig. 3.13 [35–37].

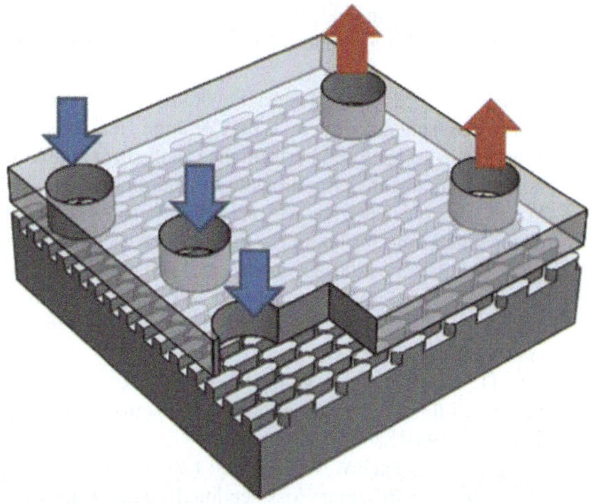

Fig. 3.13 Staggered fin micro-channel design [35]

References

1. Khamkar N, Waghmode A, Joshi A, Supekar P, Kumar A, Londhe K, Kaushik VR (2017) Heat sink design for optimal performance of compact electronic appliances—a review. J Adv Res Appl Sci 4(5):13–21
2. Mousavi H, Darzi AA, Farhadi M, Omidi M (2018) A novel heat sink design with interrupted, staggered and capped fins. Int J Therm Sci 127:312–320
3. Ahmadi M, Mostafavi G, Bahrami M (2014) Natural convection from rectangular interrupted fins. Int J Therm Sci 82:62–71
4. Haghighi SS, Goshayeshi HR, Safaei MR (2018) Natural convection heat transfer enhancement in new designs of plate-fin based heat sinks. Int J Heat Mass Transfer 125:640–647
5. Shadlaghani A, Tavakoli MR, Farzaneh M, Salimpour MR (2016) Optimization of triangular fins with/without longitudinal perforate for thermal performance enhancement. J Mech Sci Technol 30(4):1903–1910
6. Singh H, Singh H, Sehgal SS (2016) Computational analysis on performance of heat sink with different configurations of fins array. Indian J Sci Technol 9(36):1–4
7. Vyas A, Gupta S, Gupta S (2016) Determining relation among shape of perforation and convective heat transfer from lateral fin arrangement using simulation by computational fluid dynamics. Int Refereed J Eng Sci 5(4):24–31
8. Khan AA, Kim S, Kim K (2016) Performance analysis of a micro-channel heat sink with various rib configurations. J Thermophys Heat Transfer 40(3):1–9
9. Chai L, Xia GD, Wang HS (2016) Laminar flow and heat transfer characteristics of interrupted micro-channel heat sink with ribs in the transverse microchambers. Int J Therm Sci 110:1–11
10. Chai L, Xia GD, Wang HS (2016) Numerical study of laminar flow and heat transfer in micro-channel heat sink with offset ribs on sidewalls. Appl Therm Eng 92:32–41
11. Xia G, Ma D, Zhai Y, Li Y, Liu R, Du M (2015) Experimental and numerical study of fluid flow and heat transfer characteristics in micro-channel heat sink with complex structure. Energy Convers Manage 105:848–857
12. Khan AA, Kim K (2016) Evaluation of various channel shapes of a micro-channel heat sink. Int J Air-Conditioning Refrig 24(3):1–9
13. Vinoth R, Kumar DS (2018) Experimental investigation on heat transfer characteristics of an oblique finned microchannel heat sink with different channel cross sections. Heat Mass Transfer 54(1):1–9

14. Osanloo B, Mohammadi-Ahmar A, Solati A, Baghani M (2016) Performance enhancement of the double-layered micro-channel heat sink by use of tapered channels. Appl Therm Eng 102:1345–1354
15. Wong K, Ang M (2016) Thermal hydraulic performance of a double-layer micro-channel heat sink with channel contraction. Int Commun Heat Mass Transfer 81:1–7
16. Ansari D, Kim K (2017) Performance analysis of double-layer micro-channel heat sinks under non-uniform heating conditions with random hotspots. Micromachines 8(2):1–20
17. Wei X, Joshi Y, Patterson MK (2007) Experimental and numerical study of a stacked micro-channel heat sink for liquid cooling of microelectronic devices. J Heat Transfer 129:1432
18. Qu W, Mudawar I (2002) Experimental and numerical study of pressure drop and heat transfer in a single-phase microchannel heat sink. Int J Heat Mass Transfer 45:2549–2565
19. Jing D, He L (2018) Thermal characteristics of staggered double-layer micro-channel heat sink. Entropy 20(7):537
20. Jing D, Song S, Pan Y, Wang X (2018) Size dependences of hydraulic resistance and heat transfer of fluid flow in elliptical micro-channel heat sinks with boundary slip. Int J Heat Mass Transfer 119:647–653
21. Deng D, Pi G, Zhang W, Wang P, Fu T (2019) Numerical study of double-layered micro-channel heat sinks with different cross-sectional shapes. Entropy 21(1):1–16
22. Dede EM, Joshi SN, Zhou F (2015) Topology optimization, additive layer manufacturing, and experimental testing of an air-cooled heat sink. J Mech Des 137:1–9
23. Alexandersen J, Sigmund O, Aage N (2016) Large scale three-dimensional topology optimisation of heat sinks cooled by natural convection. Int J Heat Mass Transfer 100:876–891
24. Joo Y, Lee I, Kim SJ (2017) Topology optimization of heat sinks in natural convection considering the effect of shape dependent heat transfer coefficient. Int J Heat Mass Transfer 109:123–133
25. Maradiaga D, Damonte A, Manzo A, Haertel JHK, Engelbrecht K (2019) Design and testing of topology optimized heat sinks for a tablet. Int J Heat Mass Transfer 142:1–12
26. Zeng S, Lee PH (2019) Topology optimization of liquid-cooled micro-channel heat sinks: an experimental and numerical study. Int J Heat Mass Transfer 142:1–20
27. Dhaiban HT, Hussein MA (2020) The optimal design of heat sinks: a review. J Appl Comput Mech 6:1030–1043
28. Wang XD, An B, Xu JL (2013) Optimal geometric structure for nanofluid-cooled microchannel heat sink under various constraint conditions. Energy Convers Manage 65:528–538
29. Ambreen T, Kim MH (2018) Effect of fin shape on the thermal performance of nanofluid-cooled micro pin-fin heat sinks. Int J Heat Mass Transfer 126:245–256
30. Amoako AA, Doom JJ (2018) Optimization of heat sinks in a range of configurations. In 2018 Joint thermophysics and heat transfer conference, p 2945
31. Alperen Y, Sertac C (2020) Multi objective optimization of a micro-channel heat sink through genetic algorithm. Int J Heat Mass Transfer 146:118847
32. Wang H, Chen Z, Gao J (2016) Influence of geometric parameters on flow and heat transfer performance of micro-channel heat sinks. Appl Therm Eng 107:870–879
33. Ghani IA, Kamaruzaman N, Sidik NAC (2017) Heat transfer augmentation in a micro-channel heat sink with sinusoidal cavities and rectangular ribs. Int J Heat Mass Transfer 108:1969–1981
34. Naqiuddin NH, Saw LH, Yew MC, Yusof F, Ng TC, Yew MK (2018) Overview of micro-channel design for high heat flux application. Renew Sustain Energy Rev 82:901–914
35. Priesnitz BH, MuChen Z, Gao J (2016) Influence of geometric parameters on flow and heat transfer performance of micro-channel heat sinks. Appl Therm Eng 107:870–879
36. Chen L, Yang A, Feng H, Ge Y, Xia S (2020) Constructal design progress for eight types of heat sinks. Sci China Technol Sci 1–33
37. A review on recent advances in heat transfer enhancement in the micro channel heat sinks

Chapter 4
Heat Transfer Enhancement in the Micro Channel Heat Sinks

Abstract Although the various structures of the heat sinks are useful to improve heat dissipation, nanofluids flowing through the heat sink expedite the heat transfer significantly. The relationship between the relative roughness height and the Nusselt number is taken into account in the section. The relationship between the mass flow rates of the nanofluids and the heat transfer coefficient is presented. Also, the relationship between entropy generation of the nanofluids, flow direction, and the ratio of fin length to fin interval are included in the section.

Contents

This chapter elucidates the thermal performance of miniature heat sink with various conduits and effects of dimensionless numbers.

The majority of electronic and power-related equipment is currently optimized in terms of size. The miniaturization of electrical and electronic systems requires a focus on heat exchangers and better heat dissipation activity. Micro channel heatsinks are contemporary heat exchangers with microscale fluid flowing channels. The miniaturized systems are highly compact heat exchangers with increased heat transfer rates for electronic equipment. Numerous research works were conducted to enhance the thermal characteristics of the microchannel heatsinks. They reported the experimental investigation on the heat transport and the improvement in micro-scale cooling systems. Heat transfer enhancement strategies were applied for the research of microchannels and the parameters that affected the heat transfer rate are presented.

S. Harikrishnan et al., *Thermal Performance of Nanofluids in Miniature Heat Sinks with Conduits*, Fluid Mechanics and Its Applications 131,
https://doi.org/10.1007/978-981-16-7845-5_4

4.1 Introduction

During the next few decades, conventional energy sources will be depleted. Today's Research priorities include energy saving and advancing alternative energy sources [1]. Researchers and scientists are trying to accumulate conventional energy sources in energy management concerns [2]. The use of renewable energy resources must be expanded and the efficiency of conventional energy systems enhanced in order to conserve conventional energy sources [3]. Recently, research experts are more focused on thermal energy system management. Enormous research has been undertaken to enhance energy systems efficiency. The development of energy-efficient systems using miniature is an innovative technique [4]. Miniaturization has a significant impact on heat exchanger technology, making heat exchangers more portable and economical [5]. The efficiency of a heat exchanger has a significant influence on the entire efficiency and lifespan of a thermal energy system.

A micro-channel heat sink is a novel device in the field of heat exchanger mechanism. The benefits of microchannel heat sink's extensive heat transfer surface and high compactness make it the superlative heat dissipation device for electronic cooling applications. The present high-powered, compact electronic components' heat dissipation capability affects their life and performance. The life span of an electronic device is enhanced by using an effective heat dissipation device such as microchannel heatsinks. In LED cooling, combustor, refrigeration, fuel cell systems, chemical and food sectors, and so forth, microchannel heatsinks are also used.

Mini/micro channel categorization differs from standard channel classification. The channels are classified in accordance with their hydraulic diameter. Table 4.1 indicates the categories of channels with according to dimension are the most often used categories [6, 7]. A simplified category based on the hydraulic diameter of the channel was proposed [8]. The first microchannel heatsinks for the electronic cooling application was developed in 1981. The highest thermal resistance of 0.09 °C/W was recorded for the 1 cm^2 region with heat flow 790 W/cm^2 in their research of the rectangular silicon sustenance micro-channel [9]. Since then, more research has been undertaken on microfluidics to examine and optimize thermal performance and liquid flow performance by changes to channel geometries, channel aspect ratio, materials in substrates, coolant, and channel wall surface ruggedness [10]. An Indium

Table 4.1 The classification of channels

Kandlikar and Grande [6]	Mehendale et al. [7]
Conventional channels: Dh < 43 mm	Conventional channels: Dh < 46 mm
Mini-channels: 200 μm < Dh < 3 mm	Compact passages: 1 mm < Dh < 6 mm
Micro-channels: 10 μm < Dh < 200 μm	Meso-channels: 100 μm < Dh < 1 mm
Transitional channels: 0.1 μm < Dh < 10	Micro-channels: 1 μm < Dh < 100 μm

Phosphide (InP) laser heatsink microchannel with a thermal resistance of 0.07 °C and a high-capacity diode laser array has been produced [11]. The experimental research of the binary methanol–water combination has been developed by heat transfer which flows in hydraulic (Dh) microchannels from 0.113 to 0.367 mm.

The experimental test errors are dominated by diameter measurement errors that can lead to a 20% uncertainty in PI (Poiseuille number) [12]. In this continuous PI (Poiseuille number), there is no obvious variation in the flow of strokes in microscales. A 3% change in channel width and height leads in 21% uncertainty in the friction factor [13]. Apart from measurement mistakes, probable causes of the deviation pressure decrease include the entry effect, the influence of the electric double layer, and viscous dissipation [14]. Improved thermal methods for the detection of partial thermal resistors in heat sink design of the microchannel are proposed. Figure 4.1 shows the total number of study literatures published between 1996 and 2019 on micro channels in journals.

4.2 Microchannel Heatsink

A heat sink is a heat exchange device that absorbs heat from the environment using different heat transfer phenomena and working fluids. The heat sinks' miniaturization makes it more portable and stronger. The heat sinks of the microchannel are paired with fluid microscale flow channels. These are used in limited areas to disperse massive heat flows from high-power heat sources.

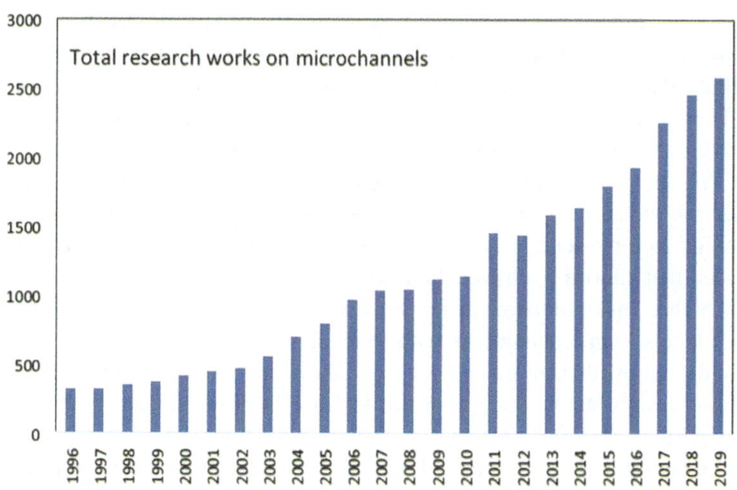

Fig. 4.1 Total number of research articles published between 1996 and 2019 on microchannels [14]

4.3 Structural Based Bifurcation

Increase of heat transfer through recirculation and vortex by the addition of microstructures and cavities along the direction of flow is a way for heat transfer improvements in the microchannel [15]. For Reynolds number (Re) of 173–635 numerical calculations are calculated on triangular chambers with the rectangular rib. The combined effects of heat and friction were explored via cavity and ribs. The research offers a means of minimizing entropy generation in order to improve heat transport. It is found that these microchannels improve when the channel regenerates more flows. In addition, this system increased mixing of fluid that contributes with an increase of Re = 500 for a rib of 0.3 and a cavity width of 2.24 to the heat transfer system.

Experiments on the various microchannel pin finishes of a maximum range of 0.0625 mm^2 are carried out [16]. Numerical research has been done of the microchannel with a sinusoidal rectangular rib cavity (Reynolds number between 100 and 800). Two metrics such as Nusselt number and friction factor were used to carry out the research [17]. According to these, due to the increase of the area, but interestingly with lower friction factor, the combined output of sinusoidal cavities with rectangular ribs is better than cavities and ribs of any other design. For the Reynolds number 800, the best results were discovered.

Numerical research is carried out on nanofluid-cooled channels. Al_2O_3, CuO and SiO_2 nano-fluids were investigated with 1–4% volume ranges below the Reynolds number of 140–1034 with a diameter of 30–60 nm [18]. Increased heat transfer was according to them owing to a volume fraction increase with reduced particle size and greatest improvement for alumina nanofluid observed. The numerical modelling of different ribs is investigated in a microchamber. The use of ribs has been shown to enhance the local heat transfer coefficient [19]. Their rectangle, triangular, diamond and ellipsoidal ribs were used in which ellipsoidal design is higher than other rib types. In comparison with conventional channels the maximum Nusselt Number was observed for the Reynolds Number (Re) has 75 with an increase of 57%.

The entropy generated by cylindrical pin structures is examined. In this research involves the creation of an ideal system with inlet speed, heat and material. The creation of entropy generation as the aspect ratio of the sink increases, and they concluded that choosing a lower fluid velocity with a lower aspect ratio combined with a higher material percentage can give the lowest entropy condition [20]. Microchannels in microchambers with triangular ribs are examined to identify the optimal design by altering their properties. They found that heat transfer rises with increasing design factors such as width and height but decreases with increasing length [21].

Pin fin configuration is examined in electronic equipment for the application of enhanced heat flux cooling [22]. They utilized a hotspot value of 2 kW/cm^2 and area of 0.25 mm^2, but 1 kW/cm^2 for the remaining zones. The improved hydrofoil fin is found to decrease the maximum temperature by 6.4 °C compared with the circular design. A micro pin fin design for the utilization of porous media in electronic cooling is improved [23]. The experiments were carried out in this case by permitting the porosity values with different pin fin positions. They found that the aforementioned two parameters were equally important for heat transfer improvement, and an ideal porosity of 0.75 with a square rib angle of 30° was proposed. A maximum Nusselt number of 24 were obtained with a pressure drop of 4.6 kPa in the instance of porous based pin fin cooling at a velocity of 1.44 m/s. With a flow velocity of 1.44 m/s, the Nusselt number rises with pressure decrease by up to 15% for a porosity of 0.56.

A microchannel with oblique fins, alumina nanofluid and a mass fraction of 0.25% were examined. There were three distinct cross-sections used: square, semicircle, and trapezoidal. The study's main purpose is to investigate flow hydrodynamics on these different designs with different mass fluxes. The trapezoidal fin obtains a 5.878% increase in heat transfer when nanofluid is used as the working fluid, which is higher than other cross-sections [24].

4.4 Double Layer Microchannel

Different strategies have been devised to reduce the temperature increase along the microchannel flow length. The experiment was carried out in a rectangular microchannel with cooling fluid flowing in the opposite direction [25]. Allowing a combined double layer mass flow rate to a single layer determines the impact of pressure drop in double layer microchannels. The double layer produced less pressure drop with an optimum pressure ratio of 2.5–3.0 [26].

As shown in Fig. 4.2, two-layer microchannels is examined for hotspot cooling in electronic devices [27]. The hotspot dimension was around 0.5 × 0.5 mm in size in this study. For this hotspot two heat fluxes of 1500 and 2000 W/cm^2 have been used while a fixed stream of 1000 W/cm^2 was used for the remainder of the chip. An integrated cooling theoretical investigation is conducted in a 3D electronic device with numerous layers of microchannels as shown in Fig. 4.3. When employing microchannels, their model predicts a decline in local wall temperature with higher

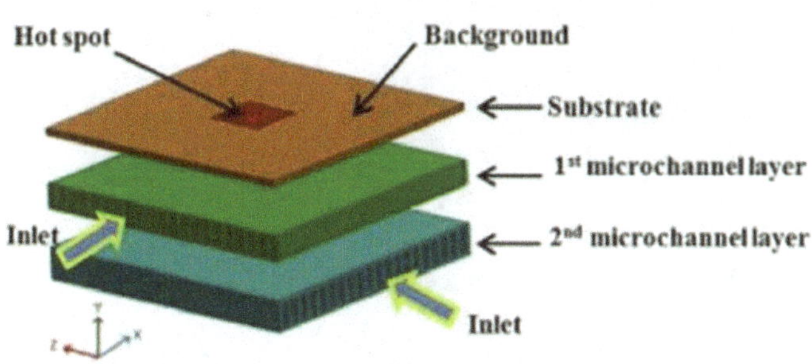

Fig. 4.2 Hotspot microchannel cooling with double layer [27]

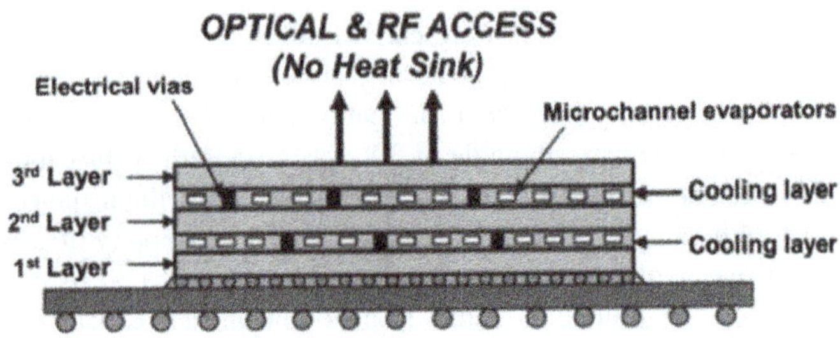

Fig. 4.3 Integrated multicooling layer microchannel [28]

exit temperature as thermal loading increases [28]. Furthermore, they are unconvinced that these predictions will hold accurate in 3D chips where loading varies with place and time; nonetheless, it is found better consistent temperature at junctions and even lower temperature inside layers. The temperature difference between layers is decreased from 15 °C to 1.5 °C [29].

In the presence of hotspots, it is found that utilizing two-phase cooling with R134 and R245 produces significant improvements than single-phase working fluids (see Fig. 4.4).

A numerical analysis of the hotspot cooling with a double layer microchannel is carried out with a rectangular cross section as shown in Fig. 4.5. The cooling of hotspots was tested in two different configurations, one central and the other evenly spread for a large number of hotspots [30]. The heat sink with parallel flow was shown to minimize the thermal resistance when a single hot spot was used, but the heatsink with transversal flow was indicated by lower thermal resistance for numerous hotspots. Comparison testing are carried out using a twin layer staggered flow channel. The microchannel of staggered type utilized in the study had a better constant temperature. It is interesting to observe that extremely reduced thermal

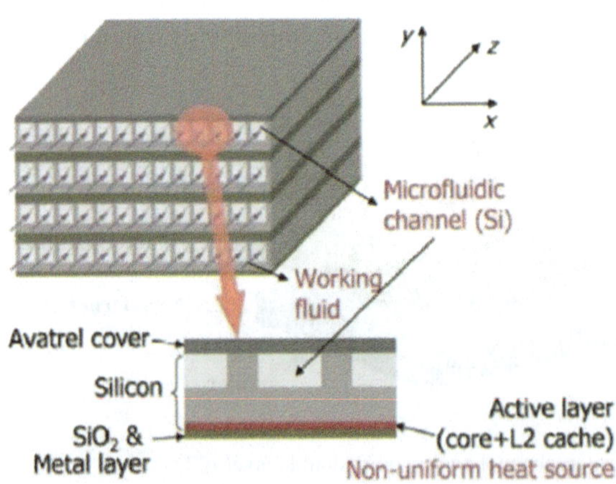

Fig. 4.4 Integrated hotspot microchannel [29]

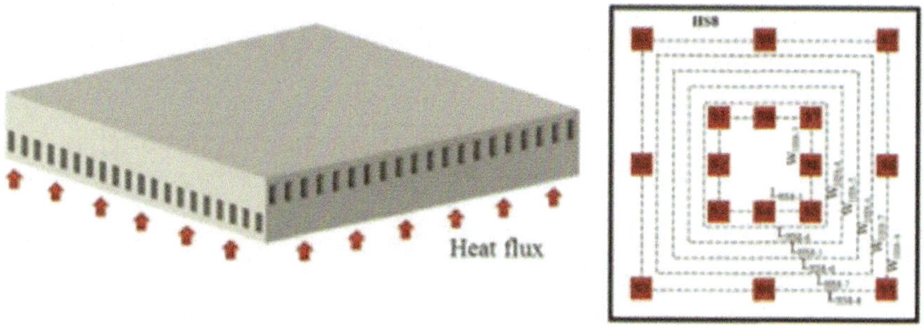

Fig. 4.5 Microchannel double layer with multiple hotspots [30]

resistance compared to a single staggered and counter-flow was produced for the same pumping-power model [31].

By introducing flux contractions in channels, numerical tests are carried out utilizing the double layer of microchannel, as shown in Fig. 4.6 [32]. The study found that a superior performance than a typical double layer is generated by tapered structures in double layer counter flow. In addition, high pump power is generated by the usage of this taper. It is determined that the combined impacts of hydraulic thermos in the structures contained are relatively minimal in contrast to usual double layers.

However, the introduction of nanofluids enhanced the performance of heat transfer and lowered thermal resistance for a fraction of 7%. Furthermore, it is found that use of nanofluids, the overall performance of a tapered double layer was not sufficient and suggested highly favored microchannels. The impact of double-layer kerosene/MWCNT nanofluid under oscillating heat flux was investigated and as shown in Fig. 4.7 [33]. There were three various dimensional (Z) lengths of the double channels such as 1/3, 2/3 and 3/3. It is determined that Z was 1/3 met with maximum thermal performance criteria in a counter flow direction under different operating circumstances and nanofluid volume fractions.

Fig. 4.6 Microchannels of double layer with contractions [32]

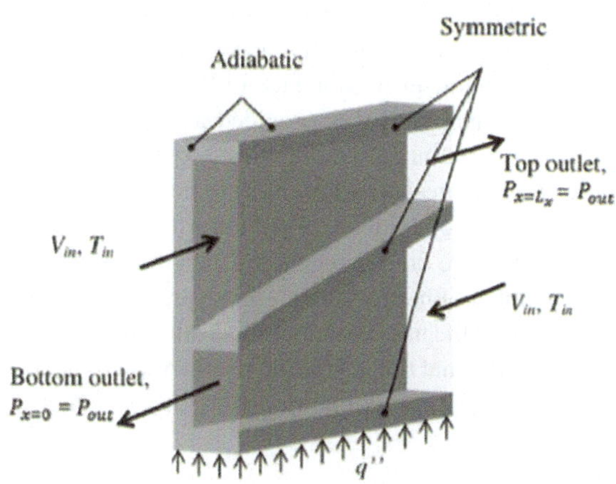

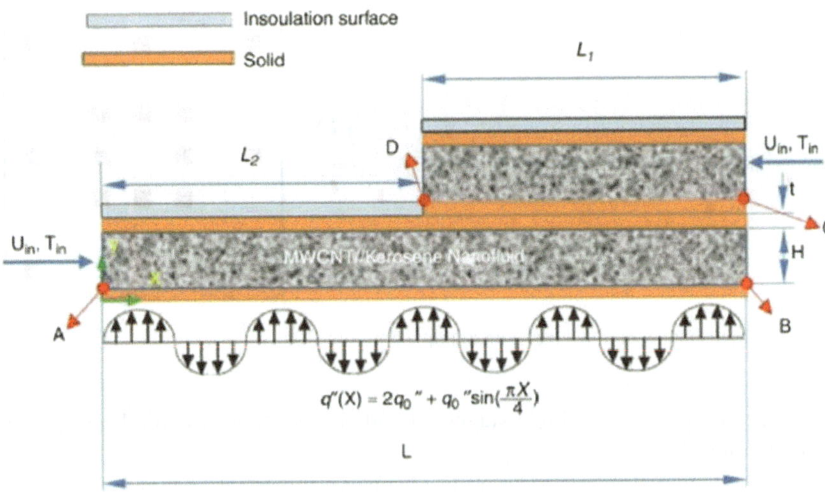

Fig. 4.7 Double layer nonlinear heat flux microchannel [33]

4.5 Effect of Microchannel Shapes

Initially, microchannel performance research works were primarily focused on determining the microchannel's useful cross-section and working fluid. Different manufacturing processes were used to provide different hydraulic diameters and aspect ratios for the microchannel. The usage of improved cross-sections with increased thermal conductive fluids significantly improves the microchannel's convective heat transfer capability. The use of these fluids substantially decreases the reaction time of cooling with a minimum amount of working fluid while increasing the heat transfer coefficient. This enhancement is due to higher Reynolds number, channel geometries, volume fraction of nanofluids and surface morphologies [34]. The result demonstrates the relationship between aspect ratio and cross section area for different wall angles in heat transfer augmentation. Copper oxide (CuO) nanofluid was employed as the basis liquid in a trapezoidal microchannel with water and ethylene glycol [35].

Higher roughness height reduces heat sink efficiency owing to reduced hydraulic diameter and raises friction factor [36, 37]. The forced convective heat transfer using aluminum nanofluid was provided and the convective coefficient of heat transfer was up to 32% increased, compared to base fluid at a mass fraction of 1.8 and with no excessive pressure drop. Furthermore, this is contradicting to the experimental data, by using circular cross-section with 0.1 mass fraction of alumina nanofluid and higher-pressure drop is registered than the base fluid [38, 39].

A number of microchannel shapes are used to analyze small heat exchangers [40]. The shape of the microchannel affects the efficiency of heat transfer and pressure loss. It is also find that the maximum results were obtained using diamond and hexagonal microchannels. By adding triangular microchannels, it was possible to increase heat transfer while also improving cooling and flow rate, resulting in a thermally stable system under high heat flux conditions [41, 42].

Surface roughness and microchannel defects have a significant impact on microchannel improvement [43]. Although it is nearly impossible to experiment with a micro fabricated device without surface irregularities, it is also extremely difficult to manufacture extremely complicated microchannel systems. However, the difference in improvements owing to surface changes may be predicted mathematically to a certain extent. Moreover, by employing an efficient numerical model for nanofluid, we can tackle practical difficulties such as agglomeration and estimate the cycle duration of the microchannel device [44].

4.6 Heat Transfer Enhancing Techniques

In the area of heat exchangers, micro-channels have a promising future. This section contains information on the methods used to enhance the microchannel heatsinks efficiency and performance.

4.7 Geometric Modifications

The main effect on the hydraulic and heat transfer performance of a microchannel is its geometry. Geometric alteration is one option for reducing pressure losses in the channel while increasing heat transfer significantly [45]. For the flow of fluids, circular and non-circular cross-sectional micro channels, with increased heat transfer areas, are used. Figure 4.8 illustrates various sectional geometries utilized

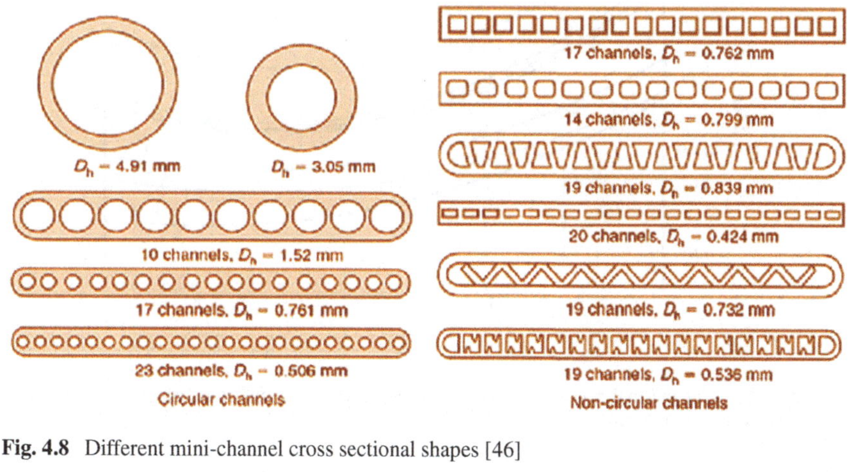

Fig. 4.8 Different mini-channel cross sectional shapes [46]

for research. Micro-channels with a trapezoidal cross section outperformed those with a rectangular cross section.

Using FC770 and Water are considered as working fluids for an experimental assessment of a rectangular hydraulic microchannel spanning between 155 and 580 μm [47]. Similar numerical investigations on rectangular microchannels found that thermal resistance decreases with constant pressure drop and increases at constant flow rate as the channel width increases. It is also observed that the friction factor initially reduces and then starts to rise when the aspect ratio increases [48]. Figure 4.9 shows the effects of the micro-channel aspect ratios (H/G) on the thermal resistance (R).

The Poiseuille number and Nusselt number are observed to rise as the side angle and aspect ratio are raised [50]. The different shapes of offset ribs on the side walls have been shown that the forward triangle ribs function best, while the rectangle ribs are less than 350 for Reynolds. Semi-circular offset ribs function well at numbers greater than 400 Reynolds. Internal fins may potentially enhance heat transfer at the cost of flow blocking pressure loss. The pressure decreases and heat transfer rates are increased, the fin height is increased and the microchannel performance as a cumulative effect is enhanced [51].

It is noticed that the heat transfer rate in microchannel increases and then decreases when the gap between the pin fins increases [52]. The microchannel fin construction with a diameter ratio of 1.0 demonstrates better thermal performance. Figure 4.10 shows schematic drawings of the numerical model utilized in this investigation. Figure 4.11 shows the implications of side wall closeness on total thermal performance.

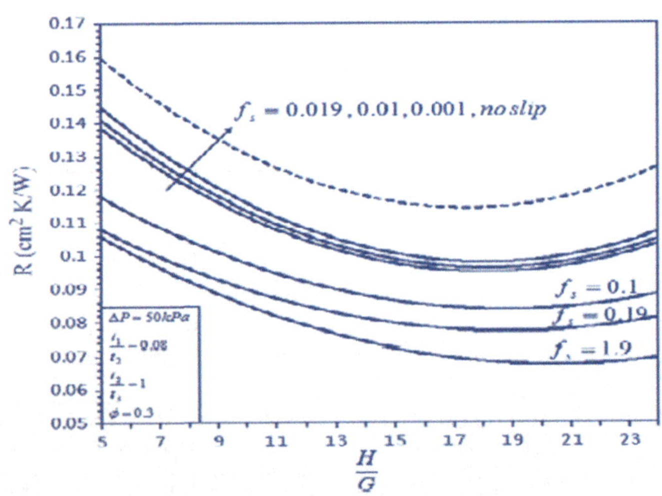

Fig. 4.9 Effects of thermal resistance (R) aspect ratios (H/G) on multiple slip coefficients when volume fraction $\phi = 0.3$ [49]

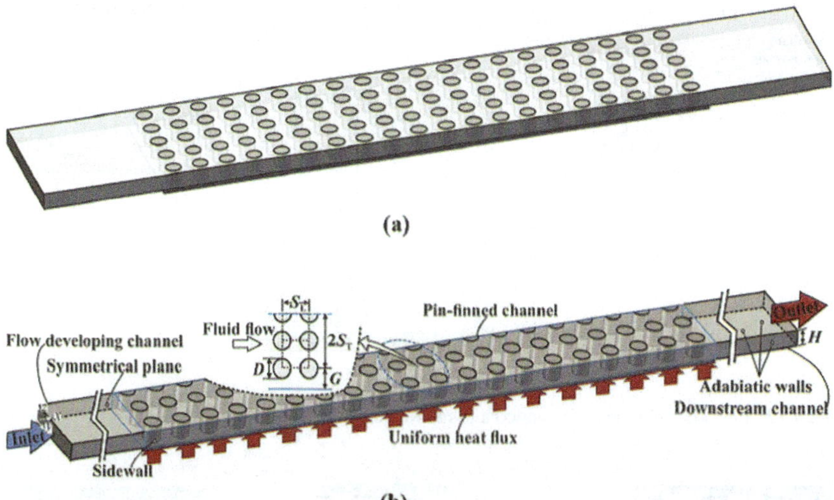

Fig. 4.10 Schematic images of the numerical model [52]

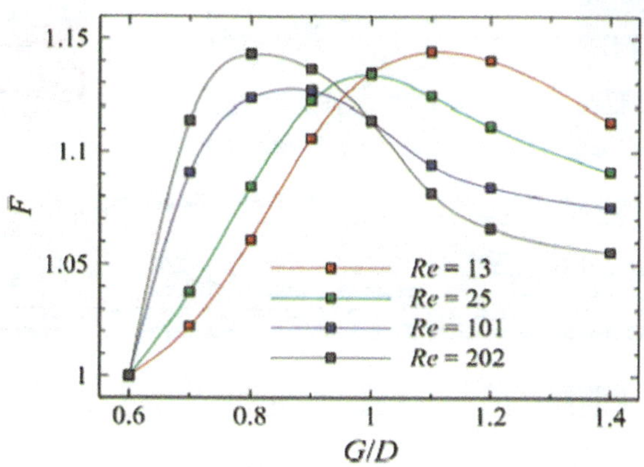

Fig. 4.11 The effects of proximity to the sidewall on thermal performance (F) [52]

Fig. 4.12 Periodic converging–diverging microchannel [53]

In a periodically convergent and divergent cross sectional microchannel (Fig. 4.12) the thermal performance was increased at the cost of pressure loss [53]. In the Reynolds range from 192.66 to 796.17 the heat transfer of converging and diverging microchannels increased by 46.8–160.2% over the direct microchannel. A periodic

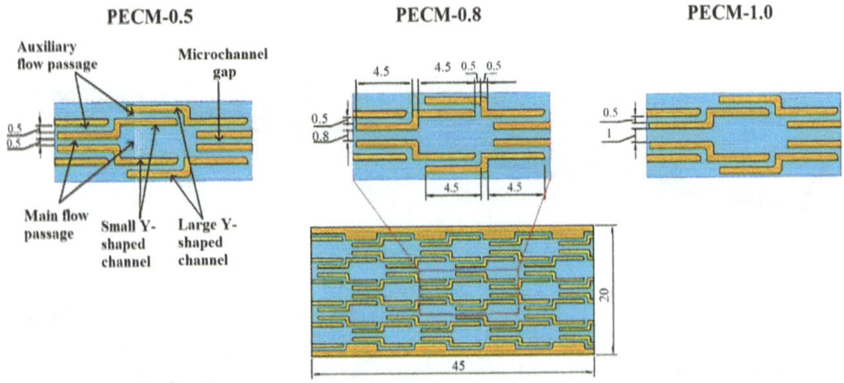

Fig. 4.13 Periodic scheme of expanded and restricted micro-channel heat sink [54]

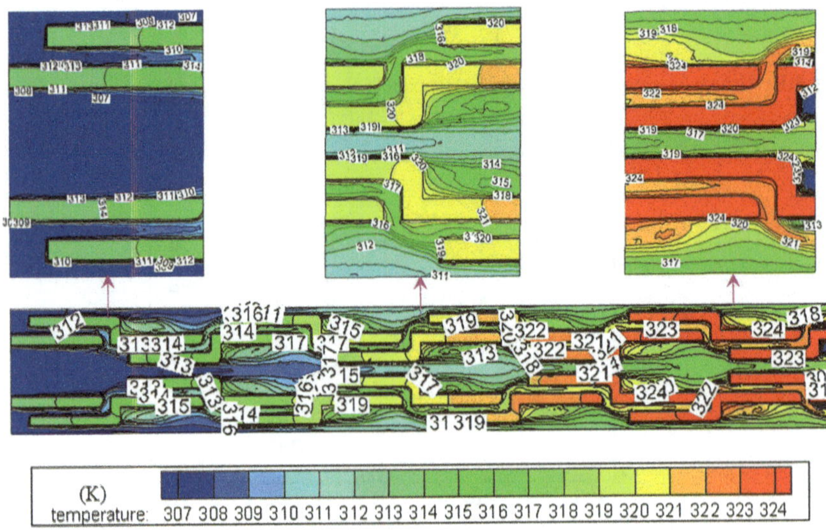

Fig. 4.14 Temperature contours of flow passages at the mid-depth plane of flow [54]

extended microchannel heatsink is created and a increase of 50–117% in the Reynolds Number (Re) range of 150–820 has been observed. Figure 4.13 illustrates the periodic expanded and restricted microchannel heatsinks design. The pressure drop and heat resistance decrease as the gap width increases, with 1.0 mm providing the best hydraulic and thermal performance. Figure 4.14 shows the temperature contours of the mid-depth flow channels.

An experimental investigation is performed on microchannel thermal sinks with periodic throttling structures (Fig. 4.15) and the average and maximum temperature of the throttled microchannel are smaller than that of the microchannel with higher pressure loss compensation [55]. Figure 4.16 shows the X-velocity contours for the fluid region in the rectangular and semicircular grooves [56].

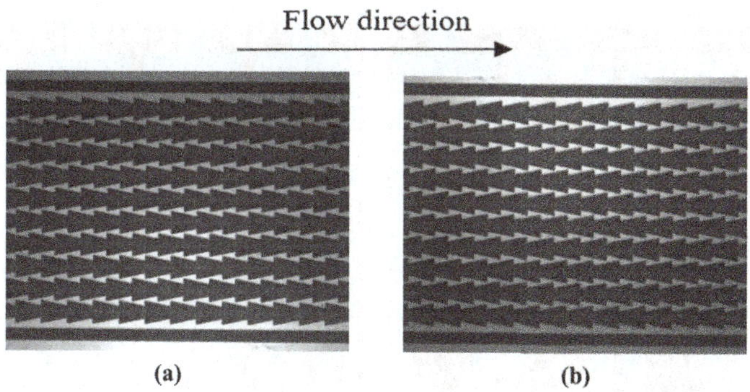

Fig. 4.15 Schematic of **a** jetting micro-channels, **b** throttling micro-channels [55]

Fig. 4.16 X-velocity contours for fluid zone in rectangular and semicircular groove [56]

It has been suggested that two different designs for the annular microchannel heatsinks (Fig. 4.17) have a better tempering uniformity in the substratum than the sequential configuration [57]. In the MRSH model, raising the slant angle results in an increase in overall heat resistance. When the slant angle is reduced from 40° to 20°, the overall thermal resistance is reduced by 9.5%. Figure 4.18 shows the average Nusselt Number (Nu) variation with pumping power.

The surface roughness of the channel wall has a substantial impact on the gas flow with a low Knudsen number (Kn). The local heat flux reduces with increasing relative roughness height in both rarefied and compressible flows. The ruggedness effect of the Poiseuille number exceeds the normal heat transfer effect. The number of Poiseuille increases as the ruggedness increases and the distance between the elements of ruggedness reduces. Figure 4.19 illustrates the average Nusselt number fluctuation with relative roughness level.

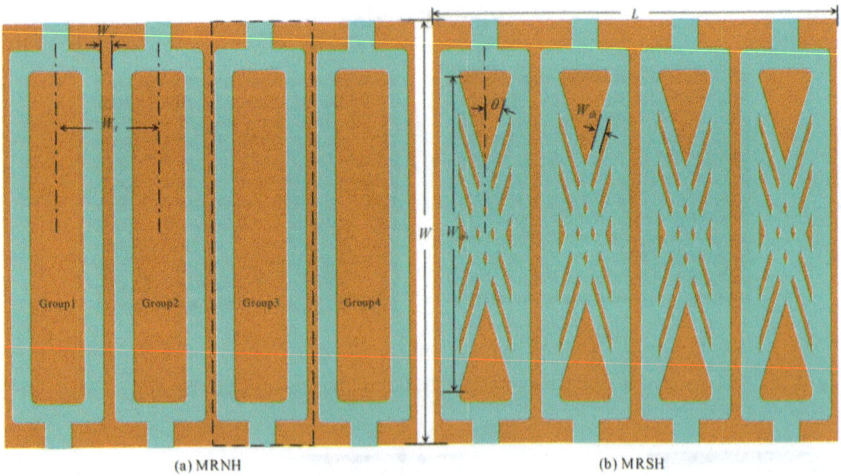

Fig. 4.17 Schematic illustrations of microchannel heatsinks, **a** MRNH configuration, **b** MRSH configuration [57]

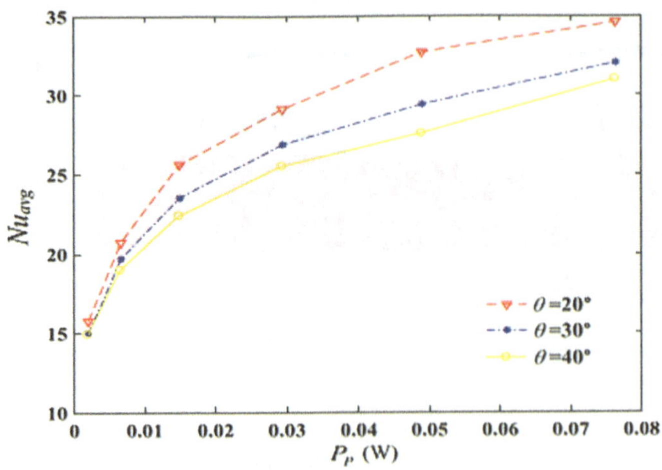

Fig. 4.18 The average Nu variation with pumping power at different slant angles [57]

The pressure drop is a significant restriction in microchannel flow. Even if the methods mentioned above improve heat transfer, an increase in pressure drop has a detrimental influence on the microchannel's behavior. A secondary flow microchannel heatsink is developed, which minimizes pressurization while increasing the efficiency of the heat transfer. The geometric design is such that the heatsink improves performance by transferring more fluid to the secondary route, redeveloping the limiting layer, and creating more vortices in secondary flow [59].

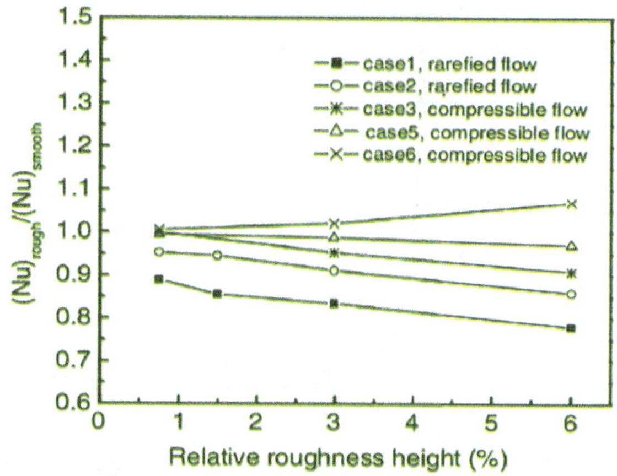

Fig. 4.19 Average Nusselt number variation with the relative roughness [58]

4.8 Nanofluids

The working fluid of a microchannel has a considerable effect on thermal and hydraulic performance. Nano fluids offer the best thermo-physical characteristics of any working fluid, making them ideal for heat transfer applications. Extensive study on microchannel heatsinks was conducted by utilizing Nano fluids as working fluids [60].

The Nusselt number increases significantly when the mass fraction increases and nanoparticles' diameter decreases. The SiO_2–water Nanofluids were observed in which exhibited best behavior of heat transfer among the four-water based nanofluids SiO_2, Al_2O_3, CuO and ZnO. Figure 4.20 shows the fluctuation in average local Nusselt number for water-based nanofluids with regard to inlet distance. The water-based CuO nanofluid fluid flow of the trapezoidal cross-sectional channel is studied and the increase in nanofluid volume has increased HT performance with more pump power and pressure drop. Figure 4.21 shows the average fluctuation of Nusselt numbers with Reynolds numbers [61, 62].

The numerical study with nanofluids on the triangular trapezoidal grooves microchannel heatsinks revealed that Al_2O_3-water nanofluid is the optimum working fluid for microchannel heatsinks-grooved trapezoidal fluids. The boiling heat transfer behavior in a swirl microchannel of an Al_2O_3–water nanofluid is susceptible to acceleration and planetary gravity [63]. The results of this experiment reveal that the size and acceleration of direction significantly affect the boiling heat transfer. The heat transfer is seen as the nanoparticles' mass fraction increases and the mass flow rate decreases, while the aspect ratio is increasing. Difference in water-based boiling heat transfer coefficients Al_2O_3 nanofluid in relation to mass flow rate is represented in Fig. 4.22.

Numeric analysis of water-based Al_2O_3 Nanofluid flow in a microchannel proves an increase in the concentration of particles in the walls and a non-uniformity of

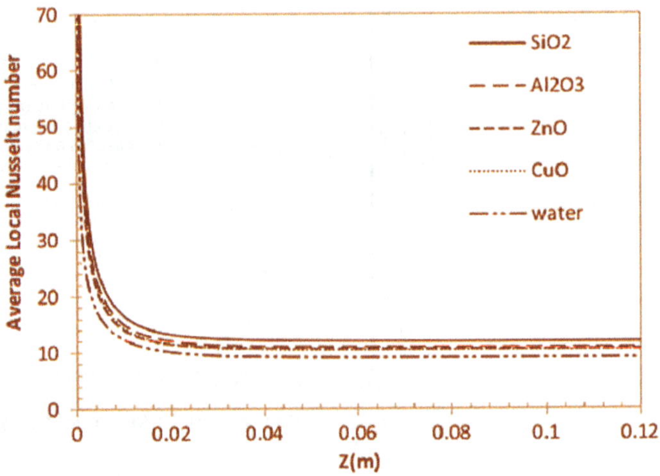

Fig. 4.20 Average local Nusselt number variation for water-based nanofluids with regard to the microchannel inlet [61]

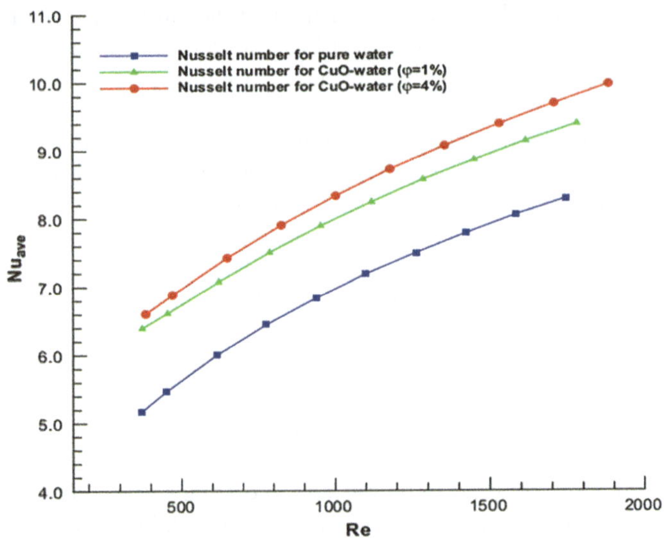

Fig. 4.21 Average Nusselt number variation of nanofluids with varying volume fractions with Reynolds [62]

the distribution in thermal conductivity with mass flow rates and particulate concentration of nanofluids [64]. Nanofluid thermal behavior improves with constant Re, although there is no linear relation between improvement and particle levels. Al_2O_3–water heat transfer coefficients Nanofluid concentrations of 1 and 2% nanoparticles are increased by 5.86% and 8.49% respectively compared to deionized water. The

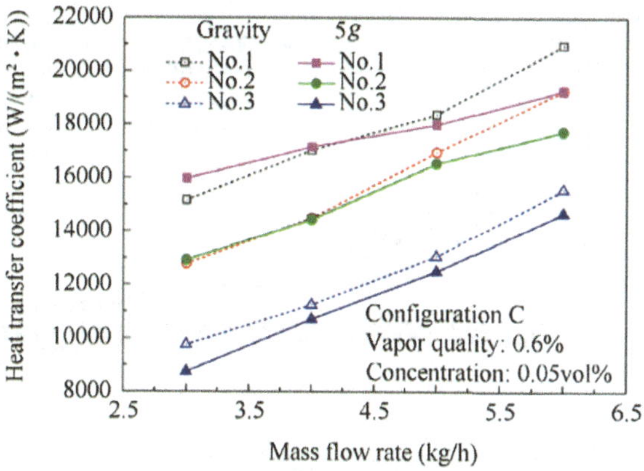

Fig. 4.22 Variation heat transfer coefficients in water-based boiling of Al_2O_3 nanofluid with varying mass flow rates for each test segment of the gravity and 5 g conditions [63]

thermal conductivity distribution at the exit as well as the axial local heat transfer coefficient is shown in Figs. 4.23 and 4.24. At different spacing in a mined micro-channel, Al_2O_3–Water Nanofluid heat transfer coefficients improve by 24.9, 27.6 and 31.1%.

The channel height influence of the water-TiO_2 nanofluid flow is explored in the rectangular heatsinks of the microchannel. It is reported that the reduction in channel height develop microchannel heat transfer, although the benefit decreases as the number of Reynolds increases. At 3 wt% concentration and Re of 200, TiO_2–water Nanofluid exhibits a maximal augmentation of 19.66% of the heat transfer coefficient. At low Reynolds numbers, the largest increase in skin friction coefficient is also found. Some studies combined the benefits of various geometries and Nano

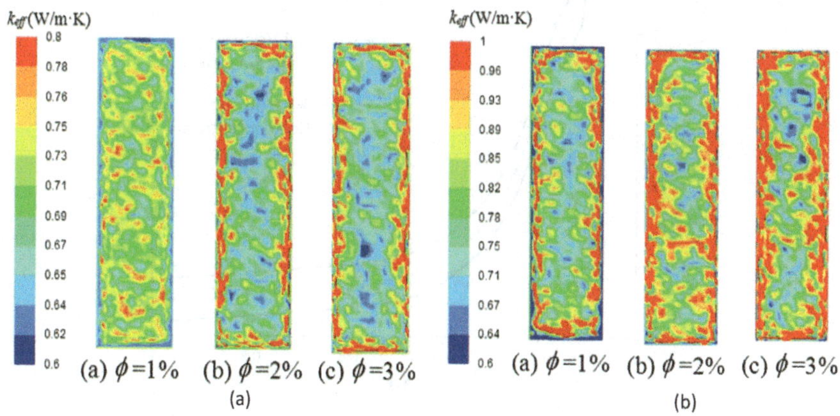

Fig. 4.23 The thermal conductivity of the nanofluids at the exit with various mass flux, **a** 0.0001 kg/s, **b** 0.0003 kg/s [64]

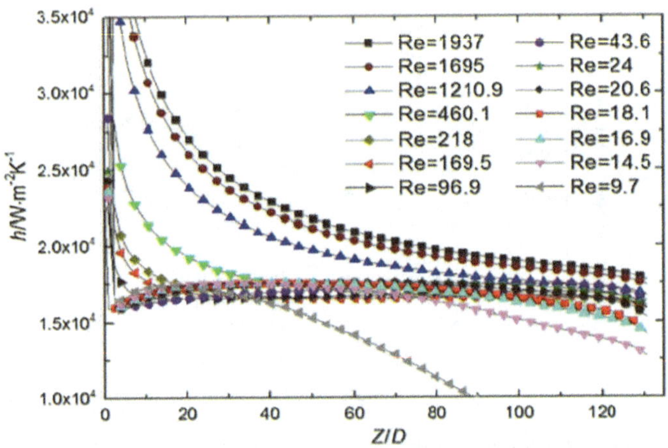

Fig. 4.24 Local heat transfer coefficient in axial direction with various Reynolds numbers [65]

fluids to increase the microchannel heatsinks heat dissipation capability. The impact of geometric changes, brownian motion and viscous dissipation on the behavior of a trapezoidal microchannel filled of nanofluid CuO–water is explored. At constant Reynolds number, raising the aspect factor (h/b) reduces the average Nusselt number and pressure loss [66].

Figure 4.25 illustrates deviations of the Poiseuille number with regard to the aspect factor [67]. It is also found that the thermophoresis effect and Brownian motion characteristics, when combined, improve heat transfer in microchannels at

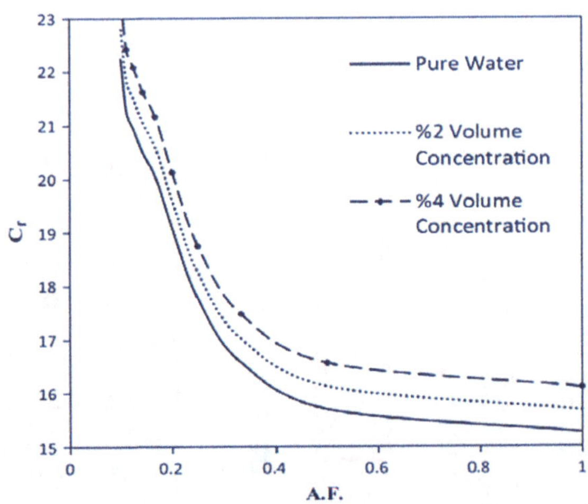

Fig. 4.25 Poiseuille number variations with aspect factor for CuO-water nanofluid volume fractions [67]

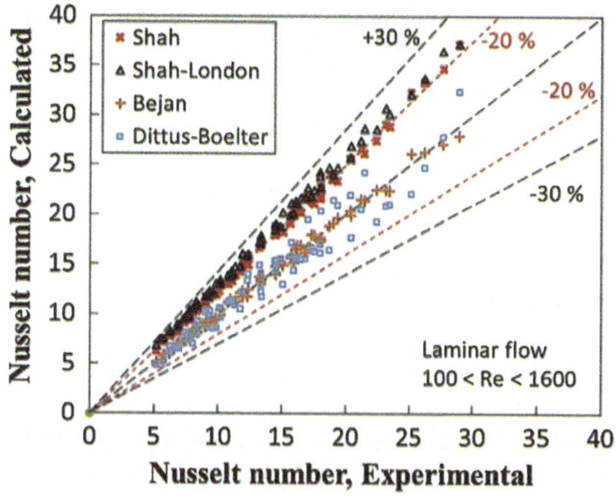

Fig. 4.26 Comparison of Nusselt number between the existing correlations and those of obtained with the experiments [68]

the expense of increased pressure loss. The Nusselt number calculated from existing correlations and experiments is compared in Fig. 4.26 [68].

Due to their improved thermo-physical characteristics, the Nanofluids exhibit enhanced heat transfer characteristics. In addition to the nanofluids benefits, there is certain negative aspect such as high pumping power demand, pressure drop and Nanoparticle sedimentation, etc. During these limits, different nanofluids have their own benefits. In the principle of hybrid nanofluids the accumulation of advantages for several nanofluids was introduced. Hybrid nanofluids with graphene-silver nanoparticles are produced in a heating system, incorporating ribs and secondary flow channels [69]. There was a rise in the heat transfer coefficient of 17%, with the concentration of nanofluid increasing from 0 to 0.1 with the number of Reynolds of 100. Figure 4.27 shows the fluctuation in the fluid and the lower surface temperature of the channel and the Reynolds number.

Entropy generation is an important factor for estimating the efficiency of an energy system. The influence of mass fraction and channel height on formation of thermal friction and entropy in a microchannel, with TiO_2 water Nano as working fluid, was explored experimentally. With increasing mass percentage, the frictional and total

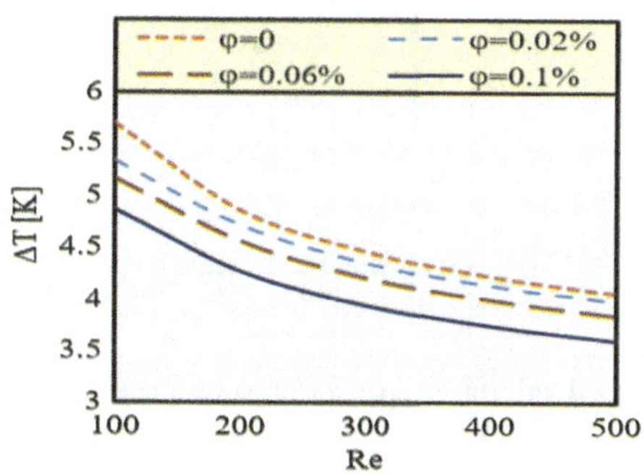

Fig. 4.27 The fluctuation of the temperature difference between channel surface and fluid at different concentration with the Reynolds number [69]

entropy creation increases while the thermal entropy generation decreases [70]. The mass fraction had the opposite effect on the creation of friction and thermal entropy at channel height. The growth in channel height increases thermal entropy creation by 30–52% and decreases channel height by 66% in the investigated channel height ranges to 98%.

With the mass fraction of TiO_2 nanoparticles increased in the studied mass fractions, frictional entropy generation increases by 3.3–21.6% and thermal entropy generation increased by 1.8–32.4%. Figure 4.28 shows a comparison of the pure water quantity in Nusselt with the prominent correlations. Microchannel heat sinks with offset strip-fin are examined and CuO nanofluid is measured using both first and second laws of thermodynamics. The solution of 0.5 wt% cellulose is used in water as a basic fluid for manufacture of CuO nanofluid. The results of the investigation demonstrated that the increases in Re improves the heat transfer performance of the heat sinking microchannels by increasing the pumping power and generating entropy and friction [71].

When the CuO Nanofluid is chosen as the working fluid, the optimum value of the heat transfer ratio improved is 2.29. Furthermore, the lowest entropy production rate of nanofluid was determined to be 2.27% lower than that of Ideal fluid. Figure 4.29 indicates the total entropy production rate of nanofluid in response to the number of fins in the direction of the flow.

This study reveals that effect of using the microchannel heat sinks and investigation carried out by various researchers has been explored.

The following are the key findings from this study.

- Low pressure drop microfluidic devices must be created since the main disadvantage of microchannel flows is the decrease pressure.

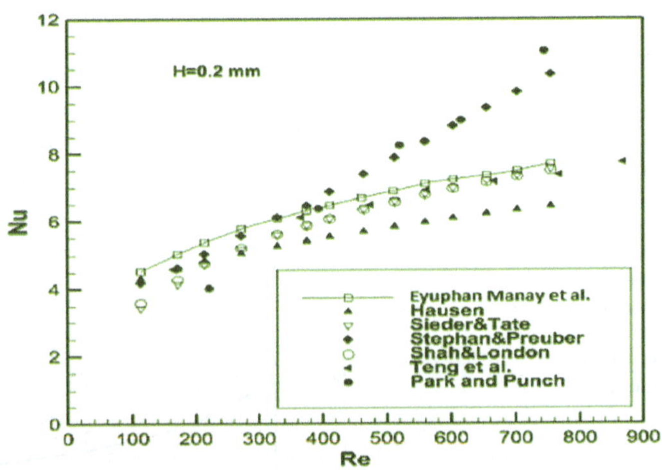

Fig. 4.28 The comparison of Nusselt number pure water with the well-known correlations [70]

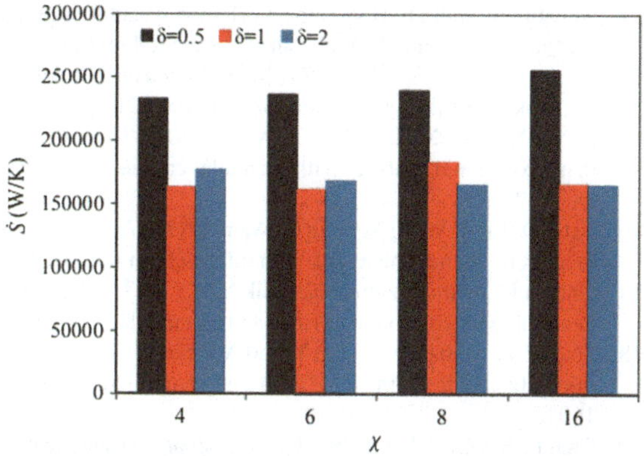

Fig. 4.29 Total rate of nanofluid entropy generation in relation to the number of fins in flow direction and ratio of fin length to fin interval [71]

- The effect of micro encapsulated phase materials on the heat transfer of a heat sink must be examined extensively.
- Current microfluid device production processes are highly costly, and the fabrication cost rises as the precision rises. A low-cost, precise manufacturing approach for microfluidic devices is still being developed.

References

1. Omer AM (2008) Energy, environment and sustainable development. Renew Sustain Energy Rev 12:2265–2300
2. Karoly R, Dumitru CD (2014) Management of a power system based on renewable energy. Procedia Technol 12:693–697
3. Nassar IA, Hossam K, Abdella MM (2019) Economic and environmental benefits of increasing the renewable energy sources in the power system. Energy Rep 5:1082–1088
4. Ameel TA, Warrington RO, Wegeng RS, Drost MK (1997) Miniaturization technologies applied to energy systems. Energy Convers Manage 38:969–982
5. Lakshminarayanan V, Sriraam N (2014) The effect of temperature on the reliability of electronic components. In: IEEE CONECCT 2014–2014 IEEE international conference on electrical, computer and communication technologies, pp 1–6
6. Kandlikar SG, Grande WJ (2002) Evolution of microchannel flow passages-thermohydraulic performance and fabrication technology. ASME Int Mech Eng Congr Expo Proc
7. Mehendale SS, Jacobi AM, Shah RK (2000) Fluid flow and heat transfer at micro- and meso-scales with application to heat exchanger design. Appl Mech Rev 53:175–193
8. Obot NT (2002) Toward a better understanding of friction and heat/mass transfer in microchannels—a literature review. Microscale Thermophys Eng 6:155–173
9. Tuckerman DB, Pease RFW (1981) High-performance heat sinking for VLSI. IEEE Electron Device Lett EDL 2:126–129
10. Missaggia LJ, Walpole JN, Liau ZL, Phillips RJ (1989) Microchannel heat sinks for two-dimensional high-power-density diode laser arrays. IEEE J Quantum Electron 25:1988–1992
11. Peng XF, Peterson GP (1996) Forced convection heat transfer of single-phase binary mixtures through microchannels. Exp Therm Fluid Sci 12:98–104
12. Judy J, Maynes D, Webb BW (2002) Characterization of frictional pressure drop for liquid flows through microchannels. Int J Heat Mass Transfer 45:3477–3489

13. Agostini B, Watel B, Bontemps A, Thonon B (2004) Liquid flow friction factor and heat transfer coefficient in small channels: an experimental investigation. Exp Therm Fluid Sci
14. Takács G, Szabó PG, Bognár G (2016) Enhanced thermal characterization method of microscale heatsink structures. Microelectron Reliab 67:21–28
15. Li YF, Xia GD, Ma DD, Jia YT, Wang J (2016) Characteristics of laminar flow and heat transfer in microchannel heatsink with triangular cavities and rectangular ribs. Int J Heat Mass Transfer 98:17–28
16. Yang D, Jin Z, Wang Y, Ding G, Wang G (2017) Heat removal capacity of laminar coolant flow in a micro channel heat sink with different pin fins. Int J Heat Mass Transfer 113:366–372
17. Ghania IA, Kamaruzaman N, Sidik NAC (2017) Heat transfer augmentation in a microchannel heat sink with sinusoidal cavities and rectangular ribs. Int J Heat Mass Transfer 108:1969–1981
18. Tokit EM, Mohammed HA, Yusoff MZ (2012) Thermal performance of optimized interrupted microchannel heat sink (IMCHS) using nanofluids. Int Commun Heat Mass Transfer 39:1595–1604
19. Chai L, Xia GD, Wang HS (2016) Laminar flow and heat transfer characteristics of interrupted microchannel heat sink with ribs in the transverse micro chambers. Int J Therm Sci 110:1–11
20. Chen L, Yang A, Xie Z, Sun F (2017) Constructal entropy generation rate minimization for cylindrical pin fin heat sinks. Int J Therm Sci 111:168–174
21. Wong KC, Lee JH (2015) Investigation of thermal performance of micro- channel heat sink with triangular ribs in the transverse micro chambers. Int Commun Heat Mass Transfer 65:103–110
22. Abdoli A, Jimenez G, Dulikravich GS (2015) Thermo-fluid analysis of micro pin-fin array cooling configurations for high heat fluxes with a hot spot. Int J Therm Sci 90:290–297
23. Zhao J, Huang S, Gong L, Huang Z (2016) Numerical study and optimizing on micro square pin-fin heat sink for electronic cooling. Appl Therm Eng 93:1347–1359
24. Vinotha R, Kumar S (2017) Channel cross section effect on heat transfer performance of oblique finned microchannel heat sink. Int Commun Heat Mass Transfer 87:270–276
25. Vafai K, Zhu L (1999) Analysis of two-layered micro-channel heat sink concept in electronic cooling. Int J Heat Mass Transfer 42:2287–2297
26. Wei X, Joshi Y (2004) Stacked microchannel heat sinks for liquid cooling of microchannel components. J Electron Packag 126:60–66
27. Abdoli A, Dulikravich GS, Vasquez G, Rastkar S (2015) Thermo-fluid stress-deformation analysis of two-layer microchannel for cooling chips with hot spots. J Electron Packag 137:031003–031008
28. Koo MJ, Im S, Joang L, Goodson KE (2005) Integrated microchannel cooling for three-dimensional electronic circuit architectures. J Heat Transfer 127:49–58
29. Kim YJ, Joshi YK, Fedorov AG, Lee YJ, Lim SK (2010) Thermal characterization of interlayer microfluidic cooling of three-dimensional integrated circuits with non-uniform heat flux. J Heat Transfer 132:41009–41019
30. Ansari D, Kim KY (2018) Hotspot analysis of double-layer microchannel heat sinks. Heat Transfer Eng 1–18. https://doi.org/10.1080/01457632.2018.1460918
31. Zhai Y, Xia G, Li Z, Wang HA (2016) Noval arrangement of staggered flow in double-layered microchannel heat sinks for microelectronic cooling. Int Commun Heat Mass Transfer 79:98–104
32. Wong KC, Ang ML (2017) Thermal hydraulic performance of a double- layer microchannel heat sink with channel contraction. Int Commun Heat Mass Transfer 81:269–275
33. Kumar A, Nath S, Bhanja D (2018) Effect of nanofluid on thermos hydraulic performance of double layer tapered microchannel heat sink used for electronic chip cooling. Numer Heat Transfer Part A Appl 73:429–445
34. Sadasivam R, Manglik RM, Jog MA (1999) Fully developed forced con- vection through trapezoidal and hexagonal ducts. Int J Heat Mass Transfer 42:4321–4331
35. Koo J, Kleinstreuer C (2005) Laminar nanofluid flow in microheat-sinks. Int J Heat Mass Transfer 48:2652–2661
36. Wu HY, Cheng P (2003) Friction factors in smooth trapezoidal silicon microchannels with different aspect ratios. Int J Heat Mass Transfer 46:2519–2525

37. Dharaiya VV, Kandlikar SG (2013) A numerical study on effects of 2D structured sinusoidal elements on fluid flow and heat transfer at microscale. Int J Heat Mass Transfer 57:190–201
38. Jung JY, Oh HS, Kwak HY (2009) Forced convective heat transfer of nano fluids in microchannels. Int J Heat Mass Transfer 52:466–472
39. Suresh S, Venkitaraj KP, Selvakumar P, Chandrasekar M (2012) Effect of Al2O3 - Cu/water hybrid nanofluid in heat transfer. Exp Therm Fluid Sci 38:54–60
40. Guo L, Xu H, Gong L (2015) Influence of wall roughness models on fluid flow and heat transfer in microchannels. Appl Therm Eng 84:399–408
41. Fani B, Kalteh M, Abbassi A (2015) Investigating the effect of Brownian motion and viscous dissipation on the nanofluid heat transfer in a trapezoidal microchannel heat sink. Adv Powder Technol 26:83–90
42. Travnicek Z, Vit T (2015) Impingement heat/mass transfer to hybrid synthetic jets and other reversible pulsating jets. Int J Heat Mass Transfer 85:473–487
43. Glazar V, Frankovic B, Trp A (2015) Experimental and numerical study of the compact heat exchanger with different microchannel shapes. Int J Refrig 51:144–153
44. Dickey JT, Lam TT (2009) High density electronic cooling triangular shaped microchannel device. US7523780
45. Tuckerman DB, Pease RFW (1981) High-performance heat sinking for VLSI. IEEE Electron Device Lett EDL 2:126–129
46. Qasem NAA, Zubair SM (2018) Compact and microchannel heat exchangers: a comprehensive review of air-side friction factor and heat transfer correlations. Energy Convers Manag 173:555–601
47. Kim B (2016) An experimental study on fully developed laminar flow and heat transfer in rectangular microchannels. Int J Heat Fluid Flow 62:224–232
48. Raj MS, Harivennkateswara R (2015) Using computational fluid dynamics and analysis of microchannel heat sink. Int J Eng Inventions 4:67–74
49. Hajmohammadi MR, Alipour P, Parsa H (2018) Microfluidic effects on the heat transfer enhancement and optimal design of microchannels heat sinks. Int J Heat Mass Transfer 126:808–815
50. Kewalramani GV, Hedau G, Saha SK, Agrawal A (2019) Empirical correlation of laminar forced convective flow in trapezoidal microchannel based on experimental and 3D numerical study. Int J Therm Sci 142:422–433
51. Chai L, Xia GD, Wang HS (2016) Numerical study of laminar flow and heat transfer in microchannel heat sink with offset ribs on sidewalls. Appl Therm Eng 92:32–41
52. Xie J, Yan H, Sundén B, Xie G (2019) The influences of side wall proximity on flow and thermal performance of a microchannel with large-row pin-fins. Int J Therm Sci 140:8–19
53. Chandra AK, Kishor K, Mishra PK, SirajAlam M (2015) Numerical simulation of heat transfer enhancement in periodic converging-diverging microchannel. Procedia Eng 127:95–101
54. Deng D, Chen L, Chen X, Pi G (2019) Heat transfer and pressure drop of a periodic expanded-constrained microchannels heat sink. Int J Heat Mass Transfer 140:678–690
55. Xia GD, Wang W, Jia YT, et al (2019) Accepted manuscript
56. Kumar P (2019) Numerical investigation of fluid flow and heat transfer in trapezoidal microchannel with groove structure. Int J Therm Sci 136:33–43
57. Liu H, Qi D, Shao X, Wang W (2019) An experimental and numerical investigation of heat transfer enhancement in annular microchannel heat sinks. Int J Therm Sci 142:106–120
58. Ji Y, Yuan K, Chung JN (2006) Numerical simulation of wall roughness on gaseous flow and heat transfer in a microchannel. Int J Heat Mass Transfer 49:1329–1339
59. Raja Kuppusamy N, Saidur R, Ghazali NNN, Mohammed HA (2014) Numerical study of thermal enhancement in micro channel heat sink with secondary flow. Int J Heat Mass Transfer 78:216–223
60. Ghale ZY, Haghshenasfard M, Esfahany MN (2015) Investigation of nanofluids heat transfer in a ribbed microchannel heat sink using single-phase and multiphase CFD models. Int Commun Heat Mass Transfer 68:122–129

61. Abdollahi A, Mohammed HA, Vanaki SM, Sharma RN (2018) Numerical investigation of fluid flow and heat transfer of nanofluids in microchannel with longitudinal fins. Ain Shams Eng J9:3411–3418
62. Li J, Kleinstreuer C (2008) Thermal performance of nanofluid flow in microchannels. Int J Heat Fluid Flow 29:1221–1232
63. Dong S, Jiang H, Xie Y et al (2019) Experimental investigation on boiling heat transfer characteristics of Al_2O_3–water nanofluids in swirl microchannels subjected to an acceleration force. Chin J Aeronaut 32:1136–1144
64. Shi X, Li S, Wei Y, Gao J (2018) Numerical investigation of laminar convective heat transfer and pressure drop of water-based Al_2O_3 nanofluids in a microchannel. Int Commun Heat Mass Transfer 90:111–120
65. Saeed M, Kim MH (2018) Heat transfer enhancement using nanofluids (Al_2O_3–H_2O) in mini-channel heatsinks. Int J Heat Mass Transfer 120:671–682
66. Martínez VA, Vasco DA, García-Herrera CM, Ortega-Aguilera R (2019) Numerical study of TiO_2-based nanofluids flow in microchannel heat sinks: effect of the Reynolds number and the microchannel height. Appl Therm Eng 161:114130
67. Fani B, Kalteh M, Abbassi A (2015) Investigating the effect of Brownian motion and viscous dissipation on the nanofluid heat transfer in a trapezoidal microchannel heat sink. Adv Powder Technol 26:83–90
68. Sarafraz MM, Yang B, Pourmehran O et al (2019) Fluid and heat transfer characteristics of aqueous graphene nanoplatelet (GNP) nanofluid in a microchannel. Int Commun Heat Mass Transfer 107:24–33
69. Bahiraei M, Jamshidmofid M, Goodarzi M (2019) Efficacy of a hybrid nanofluid in a new microchannel heat sink equipped with both secondary channels and ribs. J Mol Liq 273:88–98
70. Manay E, Akyürek EF, Sahin B (2018) Entropy generation of nanofluid flow in a microchannel heat sink. Results Phys 9:615–624
71. Al-Rashed AAAA, Shahsavar A, Entezari S et al (2019) Numerical investigation of non-Newtonian water-CMC/CuO nanofluid flow in an offset strip-fin microchannel heat sink: thermal performance and thermodynamic considerations. Appl Therm Eng 155:247–258

Chapter 5
Applications of Miniature Heat Sink

Abstract Heat transfer rates in the miniature heat sinks are found to be higher when compared to the conventional heat exchangers owing to the high surface-to-volume ratio. A direct air-cooling system using the compact cooling technique creates a huge demand for a wide range of engineering applications. Thermoelectric modules and thermo-pipes solutions are attracted due to their high heat transfer coefficients. Some of the important applications of microchannel heat sinks like electronic cooling, solar cooling, refrigeration and other specialized cooling, biomedical systems, space and defense, and other applications are included in the section.

Contents

This chapter deals with various applications of conduits miniature heat sink, suggestions for future scope of research by utilizing miniature heat sinks.

The evaporation and condensation of working fluid in a cold location can result in two phases of heat transfer which can cause the highest degree in hot fluids when compared with the conventional forced air cooling. This explains why important research concentrated on these electronic thermal control techniques. In a pool or in the direct cooler with droplets, jets or dielectric liquid sprays, electronics are immersed.

Thermoelectric coolers (TECs) are integrated with air cooling and liquid cooling techniques between cooling strategies for applications of electronic equipment. The heat pump is a solid, thermoelectric module, because the trajectory of the thermo pump is fully reversible. When a DC current passes to thermoelectric pairs, it creates one surface is cooled and the other heat surface depends on polarity [1].

The depletion of fossil fuel source of energy has become the cruise of its energy conservation techniques. Increasing the efficiency of the device is one way of ensuring energy sustainability that has been effectively implemented by miniaturization heat sink. The applications of heat exchangers for micro and mini channels in various

S. Harikrishnan et al., *Thermal Performance of Nanofluids in Miniature Heat Sinks with Conduits*, Fluid Mechanics and Its Applications 131,
https://doi.org/10.1007/978-981-16-7845-5_5

sectors have shown their promising future. In terms of smaller quantity of necessary coolant and higher than any device performance, the benefit of micro channel heat exchangers is added for these systems. However, this shows a weakness in manufacturing costs and lack of "precise performance prediction methods". Nevertheless, the applicability of micro and mini heat exchangers in cooling and air-conditioning systems has increased considerably. In particular, air-conditioning systems, R134a systems for mobile air conditioning as well as ammonia and chillers, cooling systems for steam and heat pump systems were used.

The heat exchanger will become part of the fuel cell systems on micro and small channels gradually. Convective cooling for electronic components and their systems was possible with the help of these interchangers. In micro heat exchangers in different process sectors, it can find several technologies and applications. The compactness and reduced size and weight as well as improved exchange efficiency have made these preferred decisions for cryogenic systems and space applications [2].

The electronics industry is one of the key industries that require a high-quality heat sink. Currently, high-rate cooling is crucial in electronics through increased processing speed and miniaturization of electronic components. Up to 150 W/cm^2 can be generated by electronic components. Air flows with poor refrigeration capacity for electrical elements, while in liquids with higher heat conductivity than gases there is greater refrigeration capability. Improving thermal properties in liquids leads to better cooling and it is also possible to utilize smaller electronic equipment.

Adequate heat transfer properties for nanofluids can also be accomplished with silver nanoparticles which have high thermal conductivity. Low nanofluid level prevents viscosity growth and thus prevents an increased pressure drop. Moreover, the global entropy generation (integrated) was assessed in a few studies which carried out second law analyzes of nanofluids in electric cooling [3].

High density of energy and miniaturization of current electronic equipment requires efficient solutions to prevent heat flows. The liquid refrigerated electronic systems were compact and effective because of the increased heat-conducting capabilities of liquids in relation to air. It is found that triangular pin thermal heat sinks were more effective than other pin–fin forms of heat sink [4].

5.1 Potential Engineering Applications of Nanofluids in the Miniature Channels

5.1.1 Heat Exchanger

Heat exchangers are experiencing a transformation and the recent aspiration is that heat diversions for thermal management are lightweight, compact and more effective. It is found that an increase in the input temperature from 50 to 70 °C and Reynolds number (Re) 2500, and an increase in nanoplate concentration and flow rate can

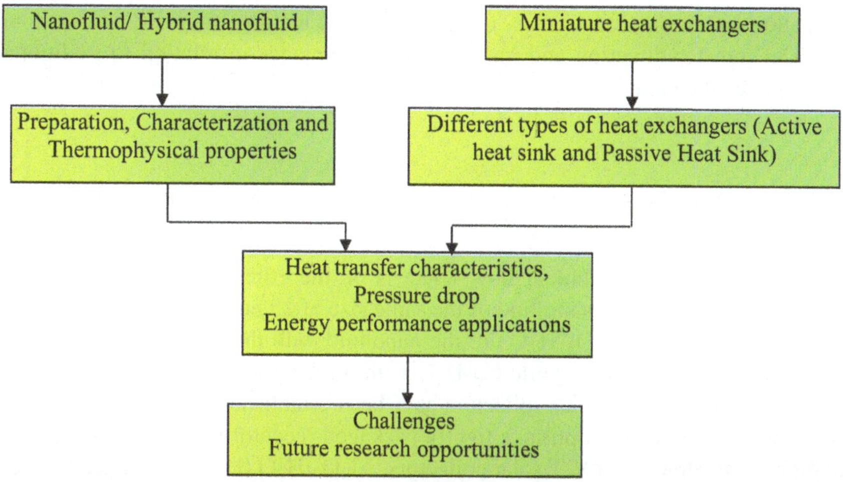

Fig. 5.1 Flowchart of miniature heat sink act as bridge of nanoparticle and various engineering applications [5]

improve the heat transfer coefficient (HTC). The efficiency of the compact heat exchanger for three different nanofluids of 0.5 vol% (Al_2O_3, CuO_2, SiO_2-ETG-H_2O) produced an 11 and 4.85% increased convective and total heat transfer coefficient.

Figure 5.1 shows the important of selection miniature heat sink and nanofluid preparation for various engineering application with consideration of heat transfer characteristics. Nanofluid transfer and pressure drop characteristics are examined for the tube and tube exchangers with both straight and spinal tubes as well as various engineering applications, including electricity production, cooling and conditioning, renewable energy, home cooling and heating. The majority of research work shows that employing nanofluids improves the heat transfer and efficiency for miniature heat exchangers [5].

5.1.2 Electronic Cooling

Due to easy manufacture, low cost and heat dissipation reliability, heat sinks are considered as a form of heat exchangers used in electronic cooling equipment and they have extended surfaces, which are flat fins or fins. It prefers the use of a fluid block for electronic thermal management with varying aspect ratio and number of channels. The increase in aspect ratio of the surface area will reduce thermal resistance for the effective cooling of the heat sink in electronic devices, which means that the optimal design of the micro scale is essential for reliable and realistic efficiency.

While most of works are used single layer heat sinks, and remaining used a multilayer stacked porous media, where porosity ranges in terms of both pressure reduction and thermal resistance, from one layer to the next (porosity scale) along the heat transfer direction and has indicated advantages in relation to traditional single-layered channels.

5.1.3 Solar Cooling

It is evaluated that a turbulent flow effect is on the efficient flow of solar mini-channel collectors using nanofluid of the same size of 25 nm with 4 V, and divulged that Al_2O_3-H_2O and Si_2O_3-H_2O are the nanofluid with the highest, moderate and lowest HTCs respectively, while Cu-H_2O with the lowest entropy generation. The results of nano-channel solar collectors have been established.

In order to study and optimize tightly packaged receptor mounting components in high-concentration CPV/T solar collectors, Al_2O_3/H_2O (low to moderate temperatures) and Al_2O_3/Synthetic Oil (moderate to high-temperature) were employed. At consistent thermal flux conditions, the nanofluids achieved an increase of 16.5% of the heat transfer rate to 2% volume of nanofluids.

Three distinct aqueous nano-fluids; Al_2O_3, CuO, and SiC were recommended to cool solar (PV) units for enhancement since a small portion of solar radiation had been transformed into electricity with the remaining portion lost in heat due to rise in cellular temperature. SiC has the highest strength and thermal conductivity (4.8% at 4%), as opposed to the two other nanomaterials while CuO showed a higher thermal conductivity compared to Al_2O_3 but a lower stability compared to other studies.

5.1.4 Refrigeration and Other Specialized Cooling

The effect of the direct metal Laser Sintering nanoparticles concentration is on the boiling heat transfer coefficient nano-refrigerants of the micro-heat exchanger where Al_2O_3 was 96 nm of particle size. The thermal and rheological properties of the Al_2O_3/antifreeze were predicted and optimized with the methodology. As a result, the models could be predicted for >99% of changes in nanofluid viscosity and thermal conductivity. Carbon nanofibre was used in ethylene and PVP as a nano-coolant at 0.1%-1wt% and the presence of nanoparticles improved the forced heat transfer coefficient of the laminar-based nanofluid with the highest concentration of 0.642 W/m.K in volume 0.5% and temperature 40 °C.

A numerical procedure with low concentration Al_2O_3-H_2O-EG studied for thermal improvement of single Proton Exchange Membrane Fuel Cell (PEMFC). Compared with the base fluid 50:50 the thermal efficiency increased 7.3 and 4.6% for 0.5 and 0.1 vol% of Al_2O_3 consecutively (water: EG). There was an increase of heat transfer in a larger volume fraction of Al_2O_3, but an increase in pumping power of up to 0.04 W due to an increase in pressure drop. Regenerative cooling system where the coolant (for example the fuel from the engine) passes through chamber tubes to distribute heat with forced convection and thermal cracking. At supercritical pressures, Al_2O_3-kerosene nanofluids were released by a vertical upwards mini tube. With increasing concentration of NP, because of changes of the internal wall surface by NPs, the HTC decreases.

5.2 Biomedical Systems

In order to analyze microflows by means of a power-operating unit, the Fe_2O_3 uses a particle diameter of 32–100 nm to determine magnetically actuated pump at flow rates of about 30 μl/min and the power supply can be regulated with the micropump unit to precisely ~5 μl/min. However, the commercialization of this invention is hindered by safety and manufacturing costs [6–9].

Despite improvements in electronics for high-efficiency IC (integrated circuit) chips, heat removal from IC chips was mostly addressed. IC chips are more compact since now, resulting in increased heat flow in smaller areas that increase working temperature which results in lowering the efficiency of the system or life if it is not cooled properly. Even a minimum of water or dielectric fluid cooling meets certain high temperature IC chips by the use of minicannels or micro-channels. The interest in nanofluids flow through the heat sink of the microchannels is therefore increasing as the study reveals that even smaller add-ons of nanoparticles are greatly increased in the thermophysical properties of the base liquid. Nanofluids are a two-phase mixture consisting of liquid base and ultra-fine particles of less than 100 nm [10].

5.3 Space and Defense

Several military equipment and systems need heat-flow cooling in the range of MW/m^2. In this respect, it is difficult to cool with traditional fluids. Driven power arms are subject to high heat flows (> 500–1000 W/cm^2) and adequate refrigeration is vital to the provision of electronic power. In such applications, like military vehicles, submarines and powerful laser dioceses, nanotropic fluids have the potential to provide needed cooling.

For the Navy and the power generation industry, transformer cooling is necessary in order to reduce the size and weight of transformers. The increasing demand for more energy will lead to the need for large scale and cost-effective replacement and/or upgrade of transformers. The substitution of conventional processing oil with a nanofluid is also a possible option. Such refurbishments will save substantial costs. The essential heat fluid in pool boiling with nanofluids increased by the order of magnitude in comparison with the base fluid alone. This means that electronic components can increase chip power or simplify cooling requirements in areas.

5.4 Other Applications

The potential impact of the use of nanofluids on the safety, neutronic and economic efficiency of nuclear systems is being examined. There are many uses in the biomedical industry for nanofluids and nanoparticles. For example, in order to prevent the harmful impacts of typical cancer treatments, iron-based nanoparticles may be utilized as medical supplies or radiation instruments without hurting its surrounding healthy tissue.

Such particles can be directed by magnets external to the body through a tumor in the bloodstream. Nanofluids can also be used to improve the safety of surgery by generating efficient cooling in the region, improving patients' survival chance and reducing their risk of organ injury. Nanofluids can be utilized as a shield to provide a better temperature around tumors in order to eliminate cancerous cells without impacting on nearby healthy cells.

The increase in heat transfer efficiency can contribute to the quality, volume and/or cost of a product or process in unlimited cases. Nanofluids are good substitute for improving heat transfer efficiency in many of these cases.

Nanofluids can be employed for energy efficiency improvements in constructions without increasing pumping capacities. Such an implementation will save on electricity, and provide environmental advantages in heating, ventilation and air conditioning systems. Nanofluids could be used in the renewable energy industry to improve heat transfer from the solar panels to storage tanks and energy density [11–15].

Recently mini heat exchangers and microchannels have been proving extremely beneficial tools to eliminate electronic high heat flows. In the first, fluid movement of the channel and the second, there are basically two key phenomena of heat transfer between the wall of the channel and fluids.

This indicates that the smaller channel dimensions of the heat exchanger provide improved heat transfer rating. By enhancing the properties of heat transfer fluids, the heat transfer rate has been raised. The thermal conductivity and quality of the other nanoparticles promised a heat transfer rate to typical fluids, for example pure water. The processing of nano-sized particles has become easier recently in the field of nanotechnology. The low thermal properties of common fluids, such as pure water, prevent energy-efficient heat exchangers from developing. But nanoparticles have emerged as an effective candidate to boost the heat transfer as technology evolves gradually [16].

Minichannels Heat sink is a way of cooling multiple machines. As a result, there have been increases the focusing on automotive, aerospace, bio-engineering cooling and electronic cooling applications. Nanoparticles, in conjunction with conventional fluids like water, are circulated along channels to increase heat transfer phenomena in tiny channels. The resulting nanofluid is transferred through the minichannels, leading to a considerable increase in the heat transfer rate.

The other kind of heat-pump is microchannel, but the pumping power demand is high compared to mini-channels. Hydraulic channels with a diameter up to 3 mm are recommended to be more suitable for high heat flow and lower pressure decreases. Most research has been carried out in recent times on mini-channels of various nanofluids [17].

The thermal efficiency of a microchannel cooled with Cu-water nanoplane and water is analyzed numerically. It showed that particle volumes and glitching velocity at the high Reynolds number influenced the heat transfer rate. The investigation on the geometric optimization of the heat sink is based on silicone using the combined number optimization and the theory of the structure. The study showed that when the dimensionless pressure drops, maximized thermal conductance increases. The effect of the surface ruggedness upon convective heat transfer in steady-state parallel microchannel two dimension plate and single-stage laminar airflow was numerically investigated. Effective mechanisms for the heat transfer rely on the highly conductive fluid and surfaces for heat transfer with high heat transfer coefficients [18].

High heat flows are produced by miniatures on computer rapid microprocessors. An effective design of a heat sink is therefore needed in order to maximize the heat and hydraulic efficiency of the electronic equipment in a safe range of temperatures. In addition to the use of nanofluids, the mini and microchannels are used to effectively extract induced heat from electronic devices for improved thermal management.

Based on the hydraulic diameters, the microchannel heat sinks are classified. The hydraulic diameters of the micro-channel sink vary between 10 and 1000 μ. The rate of heat transfer increases with reduced microchannel hydraulic diameter. Experimentally and numerically, different forms of micro-channels were examined.

If the hydraulic diameter of the heat sink is reduced, the heat transfer rate is improved effectively. The heating sinks are experimentally investigated in order to observe the effect on heat transfer by changing the thickness of the air cooled mini-channel. The channel width-to-fin thickness ratio was one in his experiments and a steady flow rate was kept. The results showed that the corresponding value of the thermal resistance was the lowest at the highest pressure drop [19].

Nanofluids of SWCNT (Single Wall Carbon Nanotube) and MWCNT (Multiwall Carbon Nanotube) have been simulated numerically, with three-dimensional flux and heat transfer in different solid volume fractions of 0–8% for water and kerosene and different Re (Reynolds number) in silicone fractal microchannels, ranging between 1300 and 3000 [20, 21].

To maintain their efficiency, efficient cooling in electronics is needed. Alumina nanofluid has been tested and shown to increase the heat transfer in microchannels. In micro-chip cooling applications, it was also shown to be effective. Nanofluid is a new concept design for a micro-channel heat sink containing cooler and the results indicated that both the thermal resistance and the temperature gradient of the warmed micro-channel walls and the coolant decreased. A cooler designed with a micro-channel sink has been employed in different concentrations to increase the hot transfer of water-Al_2O_3 nanofluids [22].

The microchannel benefits of water that contains tiny concentrated nanoparticles have recently become more attractive to nanotechnology. The coefficient of heat transfer in the microchannel has been found to be increased by nanoparticles distributed in water. The nanofluencers are effective only when the irreversibility of heat transfer is dominant across small channels under constant heat. The microchannel and nanofluid combinations provide highly leading fluids, as well as an extensive thermal area. The nanoparticles are very small and can be transferred to a solid wall. Despite all other properties, it is known as a transfer medium in the microchannel for its high thermal conductivity and improved response.

The presence of nanofluids in comparison to water has increased heat transfer in place of conventional fluids. With the increased amount of Reynolds and the concentration of particulate volume the heat transfer has increased. A certain limit of 2.0–2.5 wt% is obtained by the advantage of improved heat transfer with an increasing nanofluid particulate volume. But 5.0 wt% particle volume fraction produces the same result as pure water [23].

Microscale heat sinks are becoming critical for the current situation, heat transfer applications. The coolant is used specifically for heat removal and the generation of unit heat distribution into micro-electric systems, built-in circuit boards, laser diode frames, high energy mirrors and other compact items that pass through large parallel, machined and gravitational ducts with high transient thermal charges. The use of high-performance thermal refrigerants is complementary. The most thrilling new refrigerants is a suitable carrier liquid such as distilling water, motor oil, ethylene glycol, etc. along with small portion of nanoparticle (e.g. metals, metal oxides, or carbon-based material).

In the field of engineering heat and mass transfer, bio-medical drug delivery, aerospace technology or micro reactors were developed, starting from the automobile airbag sensors. In recent years, different silicone-based micro-systems have been developed with the advances of MEMS technology, such as microheat sinks. In the achievement of numerous technological goals involving micro and nanoenergy systems, the micro-fluid irreversibility of friction, heat transfer and electronic transport has an importance. The development of entropy covers irreversible friction and thermal properties. The generation for different slip-flow conditions, channel aspect ratios and pressure ratios in different operating parameters.

The specific purpose is to provide an even temperature distribution through the micro-electronic circuit board, integrated board, laser diode arrays, high-energy mirrors and other high-transition thermal products via a wide range of parallel, processed or grade pipe-lines. The circuit board is the very high surface-to-volume heat transfer device ratio, which results in excellent compactness and heat removal. The more fascinating new coolants are a blend of adequate fluid supplies, like distilling water, engine oil, ethylene glycol and a small portion of nanoparticle (e.g. metals, metal-oxide, or carbon based). Those liquid-particle mixtures, known as Nanofluids, showed their conductivity levels 20–150% higher than those of the basic liquids [24].

Advancing research now has dependable and effective cooling solutions for mobile devices because the market need from manufacturers is rising. To this end, the absorbed energy must be dissipated over lesser regions, leading to high heat flows. In order to cool those devices, it is most important that a fan powered heat sink is integrated, which must have a low profile for portable electronic devices. However, a major defect with this heat dissipation approach is that it needs an increased case temperature and, once this results in end-user discomfort, the portable system rapidly loses its place in this competitive market.

The requirements for heat dissipation are expected to overtake those achieved by passive mechanisms in the short term as customers need more sophisticated technological developments, and thus manufacturers are required to implement active cooling solutions. A variety of different technologies, including the incorporation of heat pipes, phase changing material, fluid cooling, or miniature fans are currently being examined for that reason. Due to a range of major developments over other technologies for mobile applications, the cooling of forced convection air using miniature fans continues to be a profit. The main benefit is that no secondary cooler is required, as there is plenty of working fluid available at ambient temperature, unlike heat pipes and fluid cooling systems. Secondly, this technology can effectively dissipate heat over longer periods, unlike phase change materials [25].

Thermal load management is an important development parameter for modern technology and industrial and scientific systems. Reduced micro and nanoelectromechanical system (MEMS and NEMS) measurements of electronic components and increased heat generation by such systems require more effective cooling methods. In particular, if the external magnet field is present in the system the use of magnetic nanofluids or ferrofluids with varied characteristics appears to boost heat transfer significantly. Ferrofluid is a colloidal, synthetic liquid or water or petroleum mixture that contains nano-particles that are magnetic for magnetism and are permanently magnetized in a single domain.

Ferrofluid has the merits for its enhanced thermal properties of normal nanoparticles. In addition, the external magnetic field is the reaction of ferrofluid. The application of a magnetic field can therefore affect the properties of ferrofluid transport [26]. The MCHS (Micro Channel Heat Sink) output was found to be improved with the decrease in the pins and/or the increase in the Reynolds number. The analysis also showed that the distribution of nanoparticles determines the temperature profile and thus affects the MCHS flow geometry. Increased heat transfer and an intensified effect in larger Re (Reynolds Number) were seen in the nanoparticles. However, the impact of nanoparticles on heat transfer at various pin sizes is complex and is influenced by the flow rate. However, a different tendency is seen in large numbers of Reynolds. For small pin sizes, the pin's increase leads to stronger action for nanoparticles before the pattern is reversed to a specific pin-size and the impact of nanoparticles decreases as the pin-sizes grow. In addition, the nanofluid viscosity has been identified as having a negative effect on the increase in heat transfer and its negative impact is greater. The effect of nanofluid viscosity again depends on the flow regime for heat transfer on different pin sizes. At low Re, the viscosity effect for small pin sizes is less, while in large numbers the pattern is reversed. Properties like size and surface power, such as nanoparticles, play an important role in improving heat transfer. If the properties of nanoparticles are disadvantageous, the nanoparticles would have practically no effect on the thermal removal process and a proper curve would therefore be applied, which allows the selectable nanoparticles to influence heat transfer most [27].

Electronic system miniaturization confronts heat rejection problems. Advances in the manufacture of these instruments have made them smaller. Such equipment generates large volumes of heat. It needs an efficient cooling system to keep the operating temperature safe. A powerful, compact and efficient cooling system, capable of sustainable durability, was needed for electronic and IT applications. The investigation of an enhanced microchannel heat exchanger was conducted on a water-cooled offset streak fin which was better than direct continuous channel walls.

Mini-fin structures have recorded the coefficient of convective heat transfer for water 9–21 times and for air 12–38 times higher than the pour plate channel. Microchannel heatsinks with trapezoidal grooved configuration are optimally designed in comparison with rectangular and triangular grooved MCHSs. In two-pass samples, the heat transfer rate was observed along with high pressure drop values. With non-uniform baffles, the mini-channel heat sink's overall thermal resistance was reduced by 9.9–13.1%. The experimental investigation of cooling efficiency in a sinusoidal wavy miniature sink, examined the effects of geometric parameters and operating fluids in comparison with a straight miniature heat sink and observed the efficient thermal power. By using a serpentine microchannel (with a square transversal) to characterize slug flow activity are examined. In critical heat flux conditions thermal protection using water can be further improved with nanofluids by the heat transfer media [28–34].

The working fluid can be divided into single and two-phase mediums within the microchannel. The new model of solar panel has significantly greater performance and lower cost compared to the conventional structure. The developed model of an air-cooling PV/T microchannel is used in heat transfer mechanism.

5.5 Application

The heat exchanger with a microchannel tube has many problems which are as follows:

(1) Problems with condensate exclusion: the condensate tube has a wide region of contact and an angle and has a rough surface and small diameter. These features prohibit the automated elimination of condensation.
(2) Surface frosting can easily build frost under low temperatures resulting to a high frequency of cleaning, due to problems excluding water on the surface of the micro channel tube.
(3) Unpredictable coolant distribution: microchannel tubes installed simultaneously. There is a distinct coolant flow rate across them. Small flow rate suggests dry flow, while the high flow rate reduces the thermal exchange potential.

The following steps should be taken at future times for achieving a low-cost and high-heat transfer rate heat exchanger [35]:

(1) Optimisation of the ends of micro-channel tubes;
(2) Improvement of the layout of micro-channel tubes;
(3) The optimization of micro-channel tubes' structure and size; and
(4) New types of material and structure developments for micro-channel tubes.

The thermal discharge system of the microchannel offers the benefits of compact structure like heights and heat exchange and also improves the efficiency of the thermal discharge compared with macroscopic heat dissipation method. In some cases more efficient coolants are needed, in spite of the substantial advances made in microchannel heat sink technology. The heat transfer efficiency of new thermal fluid can obviously be enhanced by dispersing particles with diameter of about 1–100 nm into the base liquid, and the heat conductivity is higher than the traditional liquid, hence it is also called nanofluids. However, various factors, such as solid particles (for example, size and shape), basic solutions of kind, adjunction agent, temperature and pond hydrogenous values can also be used to create thermal conductance properties for nanofluid, which is especially for SiC nanofluids and TiO_2, with a high thermal conductivity [36].

The effects of fluid flows in applications of micro scale have increased forever. The MCHS can scatter an enormous amount of heat from a small surface and can improve its efficiency with the use of nanofluids as a coolant. In MCHS, however, complex channels are found to be stronger because of their surface area, compared to simple channels and traditional coolants. For this matter, the MHSs have been proposed and validated for various configurations. Microchannel heat sinks are particularly useful for cooling electronics, and they are also the best replacement to cool such small devices. The cooling of small devices is an extremely difficult task for researchers. A microchannel in a heat exchanger is used for heat removal using nanofluids and MCHS, which is the best thermal device with high performance [37]. It was due to the viscosity increase due to the concentration increase. In addition, an increase in heat flux was demonstrated to lead to a decrease in strain. The temperature of the plate increases along with the volume. Through the use of MPCMs, the temperature of the platform decreases, indicating the higher cooling capacity of microcapsules.

The rise in Nusselt number, at the maximum of mass level in 1%, also beneficially affected the growth of microcapsules. This parameter was 9.84, 8.5 and 8.37 respectively for the Reynolds of 1000 for 1%, 2% and pure water. On the other hand, the explanation behind the 1% level was the optimum condition, namely the lack of sufficient heat flux to fully melt the 2% concentration. The Newtonian behavior of the flow with these concentrations was a study of rheological behavior of MPCMs.

Simultaneously capsulated paraffin inside nanotubes and improved its basic thermal and thermal conductivity as a working fluid for the cooling of electronic equipment. They demonstrated that the slip over the surface was improved by applying a surfactant to the water. As a result, the fluid flowing was improved and the heat transfer was improved. Due to the high latent heat of paraffin, the channel wall temperature in the CNT was reduced by roughly 3.2 °C by wax capsulation. The higher flow rate has weakened the thermal transition, because there is not enough time to complete the transformation. In addition to microchannels, bulk PCM is also employed in heat sinks.

In view of the low heat conductivity and free PCM convection, they are used as capsules (MPCMs). This means that the low thermal conductivity downside is resolved by encapsulation of this material. LMA (Low Melt Alloy) provides a high thermal conductivity thermal percolation layer leading to a decrease in thermal solution resistance. When comparing the LMA and fin uses, it was found that LMA spread the heat in all directions. Instead, the fin spread in one specific direction, LMA showing better efficiency. The stronger the heat transfer, the greater the desirable percentage of MPCMs is used.

When the saturation flow is satisfied, MPCM concentration increased and fluid film thickness declined as a result of evaporation of R-134. The results showed that the addition of 20 wt% of microcapsules decreased the temperature gradient and increased the thickness of the film through their phase changes. In reality, microcapsules absorb a portion of the heat prevented the increase in the temperature of the fuel and increased heat capacity (Peclet number).

The fluid at the end of the tube grew by 15 and 20%, appropriately with a 5–15% increase in microcapsules, and the length of the channel to the dry off increased by 29–37%. The convective coefficient was strengthened due to the lower temperature and saturation R-134 difference in the convective cells as well [38].

Heat is predictable by-product of devices and electronic circuits and an impediment to their efficiency and reliability. Thermal management is therefore crucial to improve the life of this component in modern and freshly designed electronic gadgets. The heat sink is a heat swap to remove and convert the heat flux into a refrigerating fluid (e.g. air and water) and it is also an attractive solution. It is commonly used for electronic systems, thermal control such as microchips, processors and opto-electronic products. Actually, the efficient operation of electronic devices and data centers relies heavily on reliable power and reliable refrigeration.

In reality, because to its superior thermal physical characteristics compared to basic fluids such as water and ethylene glycol, the nanofluid can accelerate heat transfer in the channels. Many tests and computer studies have been undertaken to assess the thermal and hydraulic efficiency of working fluid flow by different parameters [39]. Convective heat transfer in several techniques is one of the thermal phenomena of nanofluids. Because of the observed thermal conductivity improvements, nanofluids can have improved coefficients of convective heat transfer.

However, as nanoparticles suspension in the fundamental fluids typically alter non-thermal thermophysical parameters such as viscosity and thermal ability, it is vital to measure the impact on heat transfer efficiencies in nanoparticles. These results were obtained by measuring nanofluid's thermophysical properties: increasing thermal conductivity, increasing density, lower heat power, increasing viscosity and possible non-Newtonian behaviour. The experimental data indicate that the effects of nanofluid as a coolant in the thermal physical property and flow shape of the heat exchanger had significant roles.

The improvement of the thermal transfer capability of nanofluids provides an interesting choice in heat exchangers to improve device performance and the resulting energy efficiency advantage. On the other hand, stability of nanofluids and the cost of development of nanofluids are important factors that prevent marketing of these. In order to demonstrate the potential of nanofluids and to understand the temperature transfer properties of the Nanofluids, further theoretical and experimental research is needed into efficient thermal conductivity and viscosity and to identify new and specific applications for these areas [40].

For heat engineers, the design of the optimal microchannel is an exciting subject. Changing geometric factors such as cross-sectional shape, design, multiple and input/output port can lead to average temperature optimization, pressure drops, heat transfer coefficient, number of nutrients, flux and temperatures homogeneity. Furthermore, thermo-hydraulic efficiency, manufacturing capacity, reliability, usability and marketing must be taken into account in the design of MCHS. It was concluded that the usage of nanofluids in industrial heat exchangers, where huge quantities of nanofluids are needed, is inefficient. The findings have also demonstrated that the impact of viscosity and the low temperature decrease is important and should be taken into account before employing nano fluids in heat exchangers [41].

References

1. Putra N, Iskandar FN (2011) Application of nanofluids to a heat pipe liquid-block and the thermoelectric cooling of electronic equipment. Exp Thermal Fluid Sci 35(7):1274–1281
2. Dixit T, Ghosh I (2015) Review of micro-and mini-channel heat sinks and heat exchangers for single phase fluids. Renew Sustain Energy Rev 41:1298–1311
3. Bahiraei M, Heshmatian S (2017) Application of a novel biological nanofluid in a liquid block heat sink for cooling of an electronic processor: thermal performance and irreversibility considerations. Energy Convers Manage 149:155–167
4. Sajid MU, Ali HM (2019) Recent advances in application of nanofluids in heat transfer devices: a critical review. Renew Sustain Energy Rev 103:556–592
5. Singh SK, Sarkar J (2020) Improvement in energy performance of tubular heat exchangers using nanofluids: a review. Curr Nanosci 16(2):136–156
6. Fotowat S, Askar S, Fartaj A (2017) Experimental transient response of a minichannel heat exchanger with step flow variation. Exp Thermal Fluid Sci 89:128–139
7. Ismail M, Fotowat S, Fartaj A (2015) Simulation of Al_2O_3-ATF nanofluid in a compact heat exchanger. In: Proceedings of the 2nd International conference on fluid flow, heat and mass transfer, paper no 149, pp 1–8
8. Sidik NAC, Jamil MM, Japar WMAA, Adamu IS (2017) A review on preparation methods, stability and applications of hybrid nanofluids. Renew Sustain Energy Rev 80:1112–1122
9. Muhammad NMA, Sidik NAC, Adanta D. On the application of nanofluid in minichannel for heat transfer and fluid flow analysis. J Adv Res Des 59:11–38
10. Patel AK, Sushant B, Rajput SPS (2019) Effects of nanofluid flow in micro channel heat sink for forced convection cooling of electronics device: a numerical simulation. Int J Innov Technol Exploring Eng (IJITEE) 9(2):5230–5243
11. Yamada E, Ota T (1980) Effective thermal conductivity of dispersed materials. Wärme- und Stoffübertragung 13:27–37

12. Yang Y, Zhang ZG, Grulke EA, Anderson WB, Wu G (2005) Heat transfer properties of nanoparticle-in-fluid dispersions (nanofluids) in laminar flow. Int J Heat Mass Transfer 48(6):1107–1116
13. You SM, Kim JH, Kim KH (2003) Effect of nanoparticles on critical heat flux of water in pool boiling heat transfer. Appl Phys Lett 83:3374–3376
14. Zhang X, Gu H, Fujii M (2007) Effective thermal conductivity and thermal diffusivity of nanofluids containing spherical and cylindrical nanoparticles. Exp Thermal Fluid Sci 31(6):593–599
15. Wang XQ, Mujumdar AS (2008) A review on nanofluids-part II: experiments and applications. Braz J Chem Eng 25(4):631–648
16. Itankar MS, Walke P, Basavaraj M (2017) To study heat transfer analysis of heat sink minichannel using nanofluids. Int J Adv Res Dev 2(3)
17. Afzal A, Samee AM, Razak RA, Ramis MK (2018) Heat transfer characteristics of MWCNT nanofluid in rectangular mini channels. Int J Heat Technol 36(1):222–228
18. Ijam A, Saidur R, Ganesan P (2012) Cooling of minichannel heat sink using nanofluids. Int Commun Heat Mass Transfer 39(8):1188–1194
19. Ali M, Shoukat AA, Tariq HA, Anwar M, Ali H (2019) Header design optimization of mini-channel heat sinks using $CuO–H_2O$ and $Al_2O_3–H_2O$ nanofluids for thermal management. Arab J Sci Eng 44(12):10327–10338
20. Lyu Z, Pourfattah F, Arani AAA, Asadi A, Foong LK (2020) On the thermal performance of a fractal microchannel subjected to water and kerosene carbon nanotube nanofluid. Sci Rep 10(1):1–16
21. Azadi M, Hosseinirad E, Hormozi F, Rashidi S (2020) Second law analysis for nanofluid flow in mini-channel heat sink with finned surface: a study on fin geometries. J Therm Anal Calorim 140(4):1883–1895
22. Jama M, Singh T, Gamaleldin SM, Koc M, Samara A, Isaifan RJ, Atieh MA (2016) Critical review on nanofluids: preparation, characterization, and applications. J Nanomater 2016
23. Razali, A.A., Sadikin, A., Ibrahim, S.A. (2017) Heat transfer of Al_2O_3 nanofluids in microchannel heat sink. In: AIP conference proceedings, vol 1831, no 1. AIP Publishing LLC, p 020050
24. Li J (2008) Computational analysis of nanofluid flow in microchannels with applications to micro-heat sinks and bio-MEMS
25. Egan V, Walsh PA, Walsh E, Grimes R (2009) Thermal analysis of miniature low profile heat sinks with and without fins. J Electron Packag 131(3)
26. Ashjaee M, Goharkhah M, Khadem LA, Ahmadi R (2015) Effect of magnetic field on the forced convection heat transfer and pressure drop of a magnetic nanofluid in a miniature heat sink. Heat Mass Transfer 51(7):953–964
27. Zargartalebi M, Azaiez J (2018) Heat transfer analysis of nanofluid based microchannel heat sink. Int J Heat Mass Transfer 127:1233–1242
28. Kalteh M, Abbassi A, Saffar-Avval M, Frijns A, Darhuber A, Harting J (2012) Experimental and numerical investigation of nanofluid forced convection inside a wide microchannel heat sink. Appl Therm Eng 36:260–268
29. Tuckerman DB, Pease RFW (1981) High-performance heat sinking for VLSI. IEEE Electron Device Lett 2:126–129
30. Wang BX, Peng XF (1994) Experimental investigation on liquid forced-convection heat transfer through microchannels. Int J Heat Mass Transfer 37:73–82
31. Xie XL, Liu ZJ, He YL, Tao WQ (2009) Numerical study of laminar heat transfer and pressure drop characteristics in a water-cooled mini channel heat sink. Appl Therm Eng 29:64–74
32. Steinke ME, Kandlikar SG (2004) Review of single-phase heat transfer enhancement techniques for application in microchannels, mini-channels and microdevices. Heat Technol 22:3–11
33. Steinke ME, Kandlikar SG (2006) Single-phase liquid heat transfer in plain and enhanced micro-channels. In: Proceedings of the ICNMM2006, Fourth international conference on nanochannels, microchannels and minichannels, Limerick, Ireland, 19–21 June 2006

34. Jilte R, Ahmadi MH, Kumar R, Kalamkar V, Mosavi A (2020) Cooling performance of a novel circulatory flow concentric multi-channel heat sink with nanofluids. Nanomaterials 10(4):647
35. Zhou J, Cao X, Zhang N, Yuan Y, Zhao X, Hardy D (2020) Micro-channel heat sink: a review. J Therm Sci 29(6):1431–1462
36. Hou Y, Yang J, Zhang W (2020) Numerical study of enhanced heat transfer of microchannel heat sink with nanofluids. In: IOP Conference Series: Materials Science and Engineering, vol 721, no 1. IOP Publishing, p 012052
37. Chandravadhana A, NandaKumar V, Venkatramanan K (2020) Mono and hybrid nanofluid based heat sink technologies-a review. J Phys: Conf Ser 1644(1):012048
38. Ghoghaei MS, Mahmoudian A, Mohammadi O, Shafii MB, Mosleh HJ, Zandieh M, Ahmadi MH (2020) A review on the applications of micro-/nano-encapsulated phase change material slurry in heat transfer and thermal storage systems. J Therm Anal Calorim 1–24
39. Pourfarzad E, Ghadiri K, Behrangzade A, Ashjaee M (2018) Experimental investigation of heat transfer and pressure drop of alumina–water nano-fluid in a porous miniature heat sink. Exp Heat Transfer 31(6):495–512
40. Huminic G, Huminic A (2012) Application of nanofluids in heat exchangers: a review. Renew Sustain Energy Rev 16(8):5625–5638
41. Alihosseini Y, Targhi MZ, Heyhat MM, Ghorbani N (2020) Effect of a micro heat sink geometric design on thermo-hydraulic performance: a review. Appl Therm Eng 170:114974

The manufacturer's authorised representative in the EU is Springer Nature Customer Service Centre GmbH, Europaplatz 3, 69115 Heidelberg, Germany. If you have any concerns regarding our products, please contact ProductSafety@springernature.com

Printed and bound by CPI Group (UK) Ltd, Croydon, CR0 4YY
30/09/2026
02225307-0002